DESIGN AND ANALYSIS OF INTEGRATOR-BASED LOG-DOMAIN FILTER CIRCUITS

THE KLUWER INTERNATIONAL SERIES IN ENGINEERING AND COMPUTER SCIENCE

ANALOG CIRCUITS AND SIGNAL PROCESSING
Consulting Editor: **Mohammed Ismail**. *Ohio State University*

DESIGN AND ANALYSIS OF INTEGRATOR-BASED LOG-DOMAIN FILTER CIRCUITS

Gordon W. Roberts
McGill University

and

Vincent W. Leung
Analog Devices

KLUWER ACADEMIC PUBLISHERS
Boston / Dordrecht / London

Distributors for North, Central and South America:
Kluwer Academic Publishers
101 Philip Drive
Assinippi Park
Norwell, Massachusetts 02061 USA
Telephone (781) 871-6600
Fax (781) 871-6528
E-Mail <kluwer@wkap.com>

Distributors for all other countries:
Kluwer Academic Publishers Group
Distribution Centre
Post Office Box 322
3300 AH Dordrecht, THE NETHERLANDS
Telephone 31 78 6392 392
Fax 31 78 6546 474
E-Mail <orderdept@wkap.nl>

 Electronic Services <http://www.wkap.nl>

Library of Congress Cataloging-in-Publication Data
Roberts, Gordon W. , 1959-
  Design and analysis of integrator-based log-domain filter circuits / Gordon
W. Roberts and Vincent W. Leung.
     p. cm.
  Includes bibliographical references and index.
  ISBN: 0-7923-8699-X
  1. Log domain filters--Design and construction. 2. Electric circuit analysis.
3. Metal oxide semiconductors, Complementary--Design and construction.
  4. Bipolar integrated circuits--Design and construction. I. Leung, Vincent W.
  II. Title

TK7872.F5R63 1999
621.3815'324--dc21   99-047409

Printed on acid-free paper.
Printed in the United States of America

Table of Contents

Foreword

The bipolar transistor has a remarkable characteristic that makes it unique as a circuit design element; it displays an exponential relationship between collector current and base-to-emitter voltage that is highly accurate over an extremely wide range of currents. This conformance to a mathematical law opens up numerous possibilities for analog signal processing. Log filters represent one of the most interesting applications of this exponential relationship.

History of the Log Filter

I am often credited with being the founder of LOG filters, as I was first to publish the basic first-order circuit. However, the "real story" is somewhat more complicated. In the early 1970's, a company named dbx Inc. was founded by an engineer named David Blackmer who was famous for his intuitive insights. The dbx product line consisted mostly of a companding audio noise-reduction system that competed favorably with the Dolby-B noise-reduction system of the time, as well as numerous expander/compressor boxes to modify the dynamic range of an audio signal. The dynamic modification of an audio signal demands two basic circuit elements; a voltage-controlled amplifier to vary the gain of a signal, and an RMS level detector to sense the loudness of a signal. Dave knew that RMS detection was the only class of detectors suitable for audio gain control; like the ear, it is insensitive to the phase of the harmonics in a complex periodic signal.

In the early 1970s, the state-of-the-art for gain control was an FET used as a voltage-controlled resistor, and the most popular detector of the time was a peak detector using diodes in the feedback path of an op-amp. Neither of these approaches could meet the demands of high-quality professional audio. This is where the exponential characteristic of the bipolar transistor came to the rescue.

The first development was an improvement over the classic log-antilog voltage-controlled amplifier (VCA). Dave invented a class-AB circuit that reduced noise when the input signal was small. This VCA had a control characteristic that was exponential; that is, the gain could be expressed as;

$$\text{Gain} = \text{EXP}\left(\frac{V_c}{V_t}\right),$$

where V_c is the control voltage and $V_t = kT/q$. At room temperature this formula gives a gain control constant of about 3 mV/dB.

The next requirement for implementing a successful audio companding circuit was a true-RMS level detector. Since the VCA's had an exponential control characteristic, it was logical to try to design an RMS detector with a LOG output. That way, a detector could be connected directly to a VCA, and any processing done on the control voltage (for non-linear dynamic effects) could be expressed directly in dB. This requirement for a true RMS detector with a LOG output led to the first commercial use of a LOG filter (and indeed these detectors are still used today in many analog signal processors). The circuit came about as follows. The formula for RMS detection using an exponential "forgetting factor" leads to the following circuit

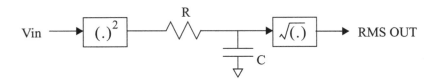

The simple R-C filter shown results in an exponential forgetting factor; more aggressive filters can be used to improve the relationship between the settling time and the amount of "ripple" present on the output voltage for sine-wave inputs.

The direct implementation of the circuit above leads to serious practical difficulties. The first approach one might take to implement this circuit is to use LOG and ANTI-LOG circuits, based on bipolar transistors, to implement the square and square-root functions. This can be done using the standard power formulas that we all learned in high-school; namely,

$$\text{LOG}(V^2) = 2 \times \text{LOG}(V),$$

and

$$\text{LOG}(\sqrt{V}) = \frac{1}{2} \times \text{LOG}(V).$$

The application of these rules leads to the following RMS circuit.

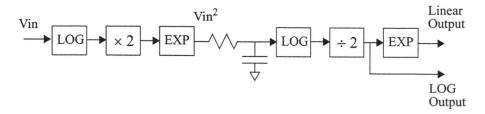

This circuit has some rather obvious problems. First, there is no way to take the LOG of a negative number. Second, the dynamic range of the squared signal is twice that of the input, leading to a serious dynamic range problem.

The solution to these problems led to the first LOG filter. The intuition provided by David Blackmer was that one could put some sort of non-linear circuit directly between the LOG and EXP blocks that would operate on the LOG of the input and eliminate the need for converting back into the linear domain before filtering.

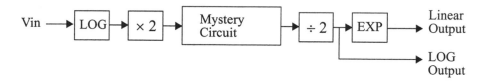

The "mystery circuit" was arrived at by reasoning that one should simply replace the resistor in the linear circuit with a diode biased at a particular quiescent current. Thus the "mystery circuit" became

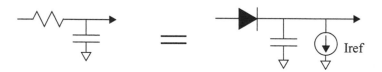

It was recognized that the "small-signal" impedance of the diode was controlled linearly by the current source, and therefore the settling time of the detector could be varied by changing the bias current.

The circuit that was first implemented looked like this;

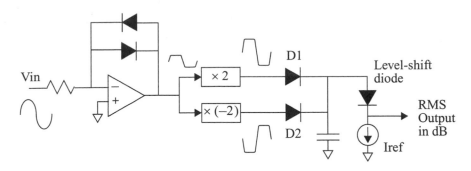

The first op-amp stage takes the LOG in two separate paths; one for positive input signals, and one for negative inputs. For positive inputs, the output of the amplifier is close to –0.6 volts, and so the input to diode D2 is close to +1.2 volts and the input to diode D1 is close to –1.2 volts. Therefore, for positive inputs, only diode D2 conducts. For negative inputs, the same argument can be applied to show that only diode D1 conducts. Thus, the "rectification" implied by the squaring operation is accomplished by modifying the original 1-diode LOG-filter circuit and replacing it with two diodes fed from opposite-polarity LOG signals. A level-shifting diode is used at the output to make the output voltage close to ground as well as to provide temperature compensation. The circuit shown here is simplified; the actual circuit uses a class-AB bias circuit that prevents the op-amp from having to make a 1.2-volt jump as the input passes through zero volts. Also, the diodes were actually made from diode-connected bipolar devices for better LOG characteristics.

When this circuit was first built, the designers did not really expect that true RMS behavior would result. After all, this "mystery circuit" seemed like a crude attempt to do some sort of filtering operation in the LOG domain. Much to their surprise, however, the circuit seemed to behave ideally. An experiment was performed where the harmonics of a complex periodic input signal were shifted in phase relative to the fundamental component. No change in output was noted, indicating ideal RMS behavior. The circuit was promptly patented and became the basis for all of dbx's analog processing equipment for the next decade.

I arrived at dbx in 1977, fresh from school, and was immediately intrigued by this circuit. All explanations for its behavior involved a lot of "hand-waving" and leaps of faith. It seemed to me that since the circuit mimicked the operation of a "true" RMS detector, then somehow the "mystery circuit" must be equivalent to the theoretical R-C filter operating in the linear domain. I analyzed the first-order LOG filter circuit and discovered that if the diode-C circuit was placed between a LOG and an ANTI-LOG converter, then from the "outside" the circuit obeyed a LINEAR first-order differential equation. In other words, even though the voltage across the "storage element" in the filter (the capacitor) was completely non-linear, the circuit could not be distinguished from a first-order RC lowpass filter.

I built a few circuits using discrete transistors and verified that LOG filter circuits seemed to indeed mimic linear filters. It was remarkable to build such a circuit and vary the diode reference current; one could observe the cutoff frequency varying by 1000:1 or more.

I attempted to extend the first-order LOG filter to higher orders, with limited success. I found that the first-order section could be cascaded to implement filters with multiple real poles, while maintaining perfect theoretical linearity. However, when I attempted to duplicate high-Q filters using Sallen-and-Key structures (by simply replacing the resistors with appropriately biased diodes), I discovered that distortion terms would arise in the equations.

I delivered a paper on this topic at the Audio Engineering Society conference in 1979, but the idea never took hold; in fact, the paper never was published in the AES journal, as the reviewer claimed it was just another form of pre-distortion (also due, I am sure, to the amateurish writing style of a green engineer!). Because of this failure to publish, the idea did not spread outside of a small circle of audio designers. In the back of my mind, however, lurked the nagging suspicion that if I was only more mathematically inclined, I might discover that LOG filters were merely one manifestation of a more general theory, and that once this theory was derived, any filter could be built in the LOG domain.

A decade or more went by, and I moved on to join Analog Devices after dbx Inc went out of business. At a conference in 1991, I had a conversation with Doug Frey, a professor at Lehigh University, whom I had known for quite some time. During this conversation I mentioned that I was still intrigued by the LOG filter concept, and that I was sure someone with better mathematical abilities than myself could make some headway into solving the higher-order filter problem. Doug became quite interested in the idea, and soon developed his theory of an exponential state-space matrix. This naturally led to a number of structures that could implement arbitrary filters with combinations of bipolar transistors, capacitors and current sources operating on a LOG version of the input signal. The publication of this work in 1993 introduced the idea to a wider audience, and spawned much of the research that has occurred over the last 7 years.

Among them, Vincent Leung, Gordon Roberts and his students at McGill University have made significant contributions to the art. They proposed the intuitive use of log-domain integrators in their LOG filter studies, a way that conventional active-RC or gm-C filters are commonly understood. As a result, interesting discoveries are made on LOG filters synthesis and analysis. I am happy to acknowledge the work published in this book.

The Future of Log Filters

As a practical circuit designer, what are LOG filters good for, and what weaknesses do they exhibit? While the elegance of the mathematics is alluring, the practical designer must compare the technique with other possible solutions. On the positive side, LOG filters are inherently scalable in both frequency and amplitude over a wide range. However, one should note that this is not always an advantage; wide tunability implies that small offsets can cause large shifts in the filter cutoff frequency. In addition, since the LOG filter concept relies on knowing the ratio between an on-chip capacitor and a transistor gm, some sort of automatic tuning loop is required if very selective filters with highly-accurate center frequencies are required. One application where the wide tuning range may be an advantage is in disk drive read channel chips. The data-rate coming from the read head varies by a wide range according to whether the head is positioned at the outside of the platter or near the center, and a filter is required that can vary its cutoff frequency over this range.

Another advantage/disadvantage is operation at low voltage and high frequencies. The signal swings within the core of a LOG filter are typically < 100 mV, and therefore operation at low voltages can be achieved. If the bias currents are reasonably high in the bipolar devices within the LOG filter core, then the impedance's are low, and thus high-frequency operation is possible. On the other hand, the low voltage swings within the LOG filter core mean that high SNR is difficult to achieve. Since the equivalent input voltage noise of a bipolar device varies inversely as the square-root of collector current (until the base-resistance dominates), it is difficult to overcome this inherently low SNR without burning a lot of power. Class-AB techniques may be used to reduce the noise when the signal is small, but the noise will then increase when the signal is large. In some applications this is acceptable, and in others it is not. In general, applications that require signal-to-noise ratios in excess of 80 to 90 dB are probably not well suited to the LOG filter design approach.

The RMS detector circuit shown in this introduction is an example of one area where LOG filters are hard to beat. If you need to raise a voltage to a power before or after a filtering operation, then LOG filters are often well suited to the task.

In terms of distortion performance, LOG filters rely on perfect LOG/ ANTI-LOG characteristics. Bipolar devices vary in their conformance to the ideal exponential law. A "good" log/anti-log circuit using bipolar devices with low bulk base and emitter resistance can probably achieve 60 dB of total harmonic distortion. We note that the errors introduced in the LOG operation can sometimes be cancelled by the same errors in the anti-log operation, but this only holds if the current densities are the same, which is unlikely if the LOG filters are to be tuned over a wide range. Another avenue for improving distortion is to use a compensation circuit that senses the current in the log/anti-log transistor and feeds back a signal to the base to correct for the intrinsic log error. Such circuit tricks have been used in commercial voltage-controlled amplifiers for quite a long time. Compensation can improve the distortion of a log/anti-log VCA to about 80 dB, and it is reasonable to expect that LOG filters will experience the same limit.

LOG filters obviously require a bipolar semiconductor process, which is sometimes hard to find in a modern mixed-signal process. Sub-threshold MOS devices can be used as LOG elements, but the currents in these devices are so low that the signal-to-noise ratio is bound to be very poor. Parasitic bipolar devices which are often found in CMOS processes are generally not useful, as the collector is normally tied to the substrate and is therefore not available as a terminal. The addition of a decent NPN to a modern CMOS sub-micron process seems to be fairly common these days, so LOG filters may still have a chance in the mixed-signal world.

The LOG filter concept has been extended to a more general class of "externally linear, internally non-linear" circuits by professor Tsividis of Columbia University. In his paper, Tsividis showed that the exponential non-linearity is only one of a general class of non-linearities that can result in an externally-linear system.

Whether or not LOG filters become commercially important, the subject seems to have spawned a number of radically new ways of thinking about linear systems that may have a greater commercial impact a few years down the road. I look forward to these future developments, and am happy to have played a role in the development of this new class of filters.

Robert Adams
Fellow
Analog Devices
Wilmington, MA
August, 1999

Preface

This book deals with the design and analysis of log-domain filter circuits. It describes several synthesis methods that aid the designer in developing bipolar or BiCMOS filter circuits with cut-off frequencies ranging from the low-kilohertz range to several hundreds of megahertz. Filter response deviations due to transistor-level nonidealities are systematically analyzed, leading to effective electronic compensation schemes. Numerous examples are provided in the text with measured experimental data from IC prototypes. This book is intended for engineers in research or development, as well as advanced level engineering students. Extensive discussion on filter test metrics should also interest test engineers who are responsible for testing high-performance, high-speed analog or mixed-signal products.

Linear analog filters are key elements in many of today's microelectronic systems. They can be found in virtually everything from cellular phones to data communication equipment to home audio entertainment centers. Of particular interest today is the integration of these filters in advanced silicon technologies, preferably CMOS or BiCMOS technologies, where analog and digital electronics can be combined on a single monolithic substrate. Nevertheless, at this time, bipolar technologies continue to play an important role in microelectronics. This is driven largely by advanced packaging technologies i.e., multi-chip modules, together with the fact that bipolar circuits can provide high-current drive capability and provide the highest levels of precision performance.

Important classes of analog filters that are required in many microelectronic systems today are continuous-time filters. Unlike the switched-capacitor technique, continuous-time filters avoid problems related to sampling and switching, such as signal settling, clock feedthrough, and charge injection. Also, they do not require any pre- or post-processing, as they are capable of handling information entirely in the analog domain. As such, they are commonly used for filtering analog signal interfacing with digital systems, e.g., anti-aliasing and reconstruction filters. The fact that these fully integrated circuits are capable of both high-speed and low-power operation makes them a popular choice for the growing wireless industry.

A log-domain filter is a novel form of a current-mode circuit that explicitly utilizes the nonlinear nature of a bipolar transistor to realize linear filter functions. Unlike conventional methods of constructing linear circuits with an *approximately* linear device, the log-domain technique makes direct use of the transistor's exponential-logarithmic behavior. Without the need for linearization, log-domain filter circuits have a very simple structure, are fast, and can operate from very low power supply levels (e.g., 1.2 V). Much research is presently underway pushing both the speed of operation and reducing the power supply level of log-domain circuits. For obvious economical reasons, efforts are also underway extending the log-domain approach to CMOS technology through the use of the MOS transistor operating in its sub-threshold region.

Robert Adams originally proposed the concept of the log-domain filter in 1979. He discovered that a diode-capacitor network, when placed between a "log" and "anti-log" converter, could realize a first-order input-output linear filter function with tunable characteristics.

. In retrospect, the log-domain technique shares certain similarities with the classical *translinear principle*. In fact, log-domain filtering can be viewed as a *dynamic* extension to this principle. As will be demonstrated throughout the text, this point of view will greatly simplify the analysis of log-domain circuits and provide useful insight into its nonideal behavior.

The practicing engineer is always looking for familiar, straightforward methods to design low-sensitivity, low-noise filters that meet required attenuation and phase requirements. Towards that goal, we shall describe in detail two different high-order log-domain filter design methods based on the internal workings of doubly terminated LC ladder networks. As will be appreciated later, these methods can be seen as an extension of classical filter design techniques. The first synthesis scheme originates from the method of operational simulation of LC ladder prototypes. The second method is based on the state-space formulation. In both cases, a simplified mathematical technique is used which eliminates much of the complicated nonlinear mathematics generally associated with log-domain circuits. We will employ this technique throughout the book.

Log-domain filter circuits suffer directly from transistor-level nonidealities. In the latter half of the book, we will study the filter response deviations due to major transistor imperfections, which include transistor parasitic emitter and base resistances, base current, Early effect, and area mismatches. Analytical equations are provided to quantify the amount of frequency response deviations caused by each parasitic element. Subsequently, compensation methods are provided. The solutions are simple and effective regardless of filter order, and, at most, involve additional current sources.

An outline of the textbook is as follows:

Chapter 1 provides the background material on conventional continuous-time integrated filters as well as the log-domain technique. We will present a historical account of the log-domain filter and its initial developments. Subsequently, the translinear principle is outlined, and circuits used to implement log-domain building blocks are derived. Next, a very simple, but yet powerful, graphical technique is described that provides the means to concisely construct a linear system from inherently nonlinear elements.

The log-domain filter synthesis and analysis methods to be discussed in this textbook are based on a single building block called the log-domain integrator. In Chapter 1 these will be described and shown to have a direct correspondence with integrators in the linear-domain. Subsequently, Chapter 2 will describe the many log-domain integrator circuits published to date. It also establishes the framework in which we can understand their input-output behavior without the need for messy mathematics.

Based on these log-domain integrators, Chapter 3 presents our first synthesis scheme for high-order log-domain filter circuits. This systematic method is a direct extension of the widely used method of operational simulation of LC ladder network prototypes. The resulting log-domain filters inherit the very low noise and sensitivity properties from their passive LC ladder prototypes. Circuit examples will be illustrated in detail based on this scheme.

However, for reasons to be outlined, the above method falls somewhat short in the creation of filter functions containing finite, non-zero transmission zeros, e.g., an elliptic response. Chapter 4 introduces our second synthesis scheme based on a state-space formulation. It is more general and can realize filter functions having arbitrary pole-zero locations. A key element of this method is its one-to-one correspondence with the state-space formulation and the resulting log-domain circuit. Design is therefore straightforward and systematic.

The filter synthesis methods presented above are exact. This means that if we assume ideal translinear elements (i.e., bipolar transistors), the resulting filter response will follow exactly the mathematically desired filter transfer function. However in reality, this is hardly the case. Log-domain circuits, in general, consist of simple circuitry. As a result, transistor parasitics have a direct effect on overall filter behavior. This leads us to the nonideality analysis section of the book.

Our analysis begins in Chapter 5, which discusses the effects of transistor nonidealities on log-domain biquadratic filters. Toward that goal, the log-domain integrator is thoroughly analyzed under parasitic emitter and base resistances, finite beta, Early effect, and area mismatches. Through our understanding of these deviation mechanisms, simple electronic compensation methods are provided.

Chapter 6 reflects on the previous nonideality study, and extends the biquadratic filter analysis to the high-order case. Based on classical *LC* ladder theories, high-order log-domain filter deviations due to transistor nonidealities are quantified. Effective electronics compensation schemes, similar to that of the biquadratic case, are proposed.

Finally, Chapter 7 describes the measured data of seven log-domain filter IC prototypes. They cover a wide range of filter orders, functions, synthesis schemes, as well as integrator structures. This chapter is intended to confirm the synthesis methods of the previous chapters, as well as to explore the limitations of log-domain circuits. It also documents and discusses the practical issues involved in experimenting with these high-speed current-mode circuits, such as the I/V and V/I interfaces, and the overall test setup. The results described here should interest those engineers and researchers who may wonder: "Well, the log-domain theory looks great on paper, but does it *really* work?"

The authors wish to gratefully acknowledge the contributions and support of numerous individuals throughout the preparation of the text. First, and foremost, the many graduate students who pioneered the work on log-domain circuit at McGill University. In particular, Douglas Perry (1993-1995) who developed the synthesis theory used throughout this textbook and provided the first experimental proof of its practicality. Mourad El-Gamal (1995-1998) who further refined the synthesis method to a fine art form, and extended the log-domain technique to the VHF range and low-voltage operation through the creation of several versatile log-domain integrator circuits. Arman Hematy (1997-1999) who extended the synthesis method to include arbitrary filter functions using the state-space formulation method. Finally, the co-author of this book, Vincent Leung (1996-1998) who developed a theory that quantifies the effects of various transistor nonidealities on the filter's frequency response. The work of these individuals all figure prominently in this book.

We would also like to acknowledge the support from the Canadian National Science and Engineering Research Council, MICRONET, a Canadian federal network of centres of excellence dealing with microelectronics devices, circuits and systems for ultra large-scale integration, and the Canadian Microelectronics Corporation. Thanks also go to Gennum and Nortel Networks Corporations for fabricating the experimental chips discussed in this book.

We would also like to extend our sincere appreciation to all the dedicated staff members and graduate students associated with the Microelectronics and Computer Systems (MACS) Laboratory at McGill University. In particular, we would like to thank system administrators Jacek Slaboszewicz, Ray Daoud and Andrew Staples for maintaining the computer infrastructure in our laboratory. We would also like to thank our administrative assistant Connie Greco for her unwavering support. The helpful discussions with graduate students John Abcarius, Benoit Dufort, Ara Hajjar, Choon Haw Leong, Loai Louis and Benoit Veillette cannot be understated. We also like to thank Katie Silverthorne for help with proofreading this textbook.

Finally, Gordon Roberts would like to extend a sincere thanks to Eileen O'Reilly and their two children, Brigid Maureen and Sean Gordon, for their continued support and encouragement on work that has taken family time away from them. Vincent Leung would like to thank his family for their unconditional love and sacrifice, and his wife Ming-Yan Venus Chiu, who has always been his true companion through all the joyful and difficult times. He would also like to express his utmost gratitude to God.

G. W. Roberts
Montreal, Quebec, Canada

V. W. Leung
Somerset, New Jersey, USA

August, 1999

CHAPTER 1   **Introduction**

Frequency shaping networks, or filters, are key elements in many of today's microelectronic systems. They can be found in everything from cellular phones to data communication equipment to home audio components. Filters generally fall into three broad categories: continuous-time, sampled-data or fully digital. Digital filters have all of the advantages associated with digital systems and can be easily incorporated inside the DSP (Digital Signal Processor) core of an integrated circuit. They are best suited for lower frequency applications and find widespread use in fully digital designs. Sampled filters combine analog filtering techniques with digital sampling principles. This makes them ideally suited for data converters (analog-to-digital or digital-to-analog) which must interface between the real analog world and the digital core of most microelectronic systems. Sampled-data systems generally use MOS technology, which allows them to be integrated on the same chip as the digital circuitry in a cost-effective manner. One popular example would be switched-capacitor filters. With their accurate frequency response, good linearity and high dynamic range, these filters have found numerous applications in integration within ASICs (Application Specific Integrated Circuits) and standard product devices.

Continuous-time filters make up a small but important part of the filter design area. They avoid problems related to sampling and switching, such as settling, clock feedthrough, and charge injection, etc. They are especially effective when dealing with real-world (analog) signals and are commonly used in high frequency, low-power systems. This last advantage makes them a popular choice for the growing wireless industry.

The two most popular forms of integrated continuous-time filters are transconductance-C (gm-C) and MOS-C filters [1]. These are well suited for integrated circuits as they lend themselves well to fully differential design and can be fabricated using only grounded capacitors. The following section gives a brief description of these two continuous-time filtering techniques.

1.1 Conventional Continuous-Time Integrated Filters

1.1.1 MOS-C Filters

A MOS-C filter can be constructed by replacing the resistor in a standard active-RC filter by a MOS transistor biased in the triode region [1]. The equation for the current flowing through such a MOS transistor would be [2]

$$i_D = \frac{1}{2}\mu C_{ox}\left(\frac{W}{L}\right)[2(V_C - V_2 - V_T)(V_1 - V_2) - (V_1 - V_2)^2] \tag{1.1}$$

where the signal symbols are defined in Figure 1-1(a). The above equation can be separated into linear and non-linear terms, such that

$$i_D = \mu C_{ox}\left(\frac{W}{L}\right)(V_C - V_T)(V_1 - V_2) + \text{a non-linear term} \tag{1.2}$$

Therefore, if we neglect the non-linear part of the equation, the MOS transistor can be used as a voltage-controlled resistor where

$$R(V_C) = \frac{1}{\mu C_{ox}\left(\frac{W}{L}\right)(V_C - V_T)} \tag{1.3}$$

Such an approach would result in the single-ended MOS-C integrator shown in Figure 1-1(b). Unfortunately, the non-linear portion of the MOS current equation will be appreciable for all but the smallest input levels. One solution is to use a fully balanced design like the one shown in Figure 1-1(c). This eliminates the even portion of the non-linear function, which accounts for most of the non-linearity (the odd portion typically accounts for less than 0.1% or -60 dB).

The design of different types of MOS-C filters would follow the same approach as is used for the design of active-RC filters [3]. Due to the voltage-controlled resistor, these types of filters are well suited to designs that incorporate some form of automatic tuning [4]-[6]. MOS-C filters typically show distortion levels in the order of 40-60 dB due to the non-linearity of the MOS transistor acting as a resistor. However, with feedback techniques, distortion levels of -90 dB have been reached [7]. Their major drawback is that they have limited frequency operation on account of the finite op amp bandwidth and are therefore not particularly well suited to high-frequency applications.

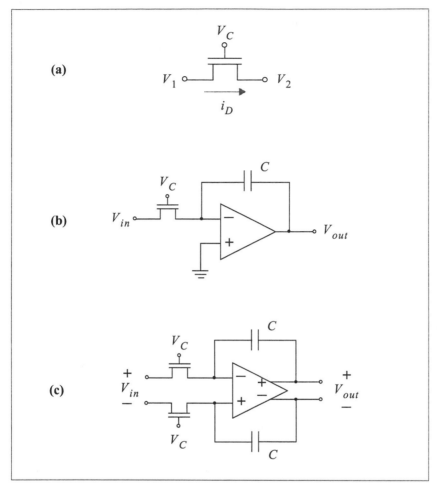

Figure 1-1: (a) MOS resistor, (b) single-ended MOS-C integrator, and (c) fully-differential MOS-C integrator.

1.1.2 Transconductance-C (g_m-C) Filters

Transconductance-C filters are based on the operational transconductance amplifier (OTA) which generates an output current that is proportional to the input voltage by a factor of g_m. The symbolic representation for an OTA is shown in Figure 1-2(a) along with a simple MOS implementation that is composed of a differential pair, three current mirrors and a simple current source (Figure 1-2(b)) [8]. The OTA can be used to build an integrator by simply pushing the output current into a capacitor as shown in Figure 1-2(c). Deriving the transfer function of the g_m-C integrator, we obtain

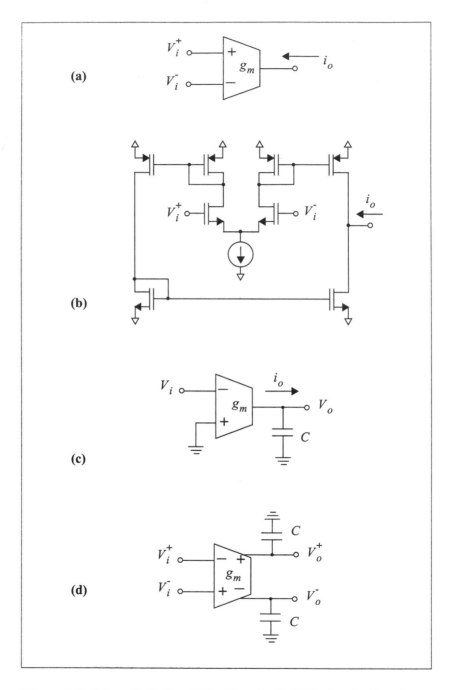

Figure 1-2: (a) symbol of an OTA, (b) a simple OTA circuit, (c) a single-ended g_m-C integrator, and (d) a balanced g_m-C integrator.

$$H(s) = \frac{V_o}{V_i}$$

$$= \frac{g_m V_i \left(\frac{1}{sC}\right)}{V_i} \tag{1.4}$$

$$= \frac{g_m}{sC}$$

In order to increase the signal-to-noise ratio by reducing the common-mode noise, for example, caused by the switching transients of the digital circuit, these filters are usually built using the fully-differential or balanced form shown in Figure 1-2(d). Given the g_m-C integrator, traditional filter design methods such as LC ladder simulation can be used to achieve different filter functions.

Transconductance-C filters are more suited to high-speed applications than the MOS-C filters described previously since they can be used in an open-loop configuration and thus need not be constrained by the stability requirement, which limit op amps. Several g_m-C filter designs have been proposed which are suitable for video-rate applications [9]-[10]. The drawback to using the OTA in an open-loop configuration is that the circuit is limited to very small input levels in order to keep it relatively linear. For example, the circuit of Figure 1-2(b) would need a differential input of less than 50 mV for reasonable results. Many different techniques have been proposed which increase the input range while maintaining linearity. These often degrade the frequency response due to added parasitics [11]. Several circuits that combine high linearity with a relatively high bandwidth can be found in the literature [12]-[13]. A final drawback concerning g_m-C filters is their dependence on the parameter g_m which makes them highly susceptible to process variations. This can be accounted for on-chip by including some form of automatic tuning [14]-[15].

In summary, MOS-C filters show good distortion behavior but suffer at high speeds due to the frequency compensation of the closed-loop amplifiers. Transconductance-C filters offer greater frequency range of operation but often at the expense of linearity. There is a constant quest for a continuous-time filtering technique that can combine high-frequency performance with low distortion levels.

Log-domain filters have recently emerged and captured tremendous research attention by showing potential to fulfil both of these stringent requirements. Log-domain filtering explicitly employs the diode nature of bipolar transistors, resulting in a class of frequency-shaping translinear circuit [16]. Promising results in high-speed, high linearity, and low-power applications have been recently demonstrated [17]-[19]. Most interesting of all, it opens the door to elegantly realizing a linear system with inherently non-linear (logarithmic-exponential) circuit building blocks, and may achieve the advantageous potential of companding (compress-expand) signal processing [20].

We will start our log-domain investigation by studying the groundbreaking work of Robert Adams in 1979.

1.2 Introducing Log-Domain Filters: Adams's 1979 Discovery

The concept of log-domain filter was originally invented by Adams and introduced to the Audio Engineering Society in 1979 [21]. He recognized that the diode-capacitor combination could be used to replace the resistor-capacitor pair for filtering. Advantageously, by controlling the bias current on the diode, the filter cutoff frequency can be electronically tuned over several decades of frequency. In the most general terms, he conceived the log-filter as: "*a circuit, composed of both linear and non-linear elements, which, when placed between a log converter and an anti-log converter (in the "log domain"), will cause the system to act as a linear filter.*"

The block diagram in Figure 1-3 can graphically illustrate this idea. The meanings of the log and anti-log converter will become obvious as we walk through the following design example.

1.2.1 An Exact First-Order Log-Domain Lowpass Filter

Adams considered the first-order linear RC lowpass filter as shown in Figure 1-4(a). It implements the differential equation,

$$V_o + RC \cdot V_o' = V_{in} \qquad (1.5)$$

where V_o and V_{in} are the output and input signals, respectively, and V_o' denotes the time derivative of V_o. Trying to duplicate this filter function in the log-domain, he suggested the log filter section of Figure 1-4(b). It is similar to its linear counterpart in the sense that the resistor is replaced by the diode D_1 biased by current I_o. The rest of the circuit can then be viewed as a buffer or level shifter to make up for the diode drop

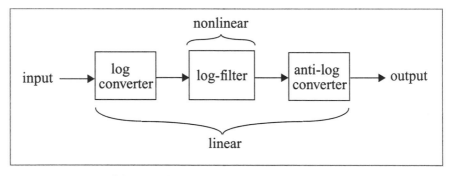

Figure 1-3: Concept of log-domain filter.

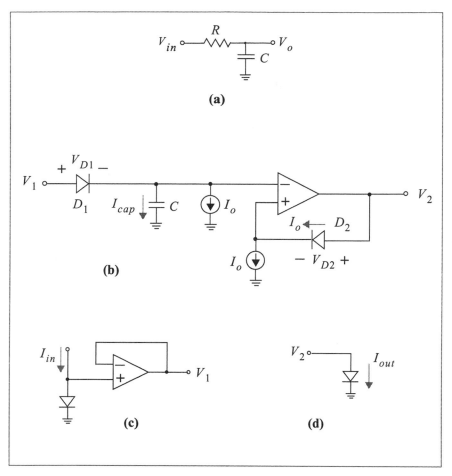

Figure 1-4: First-order log-domain lowpass filter: (a) RC prototype, (b) corresponding log-filter, (c) log converter and (d) anti-log converter.

across D_1. Writing the KCL equation at the capacitor node, we have

$$I_{cap} = I_{D1} - I_o \qquad (1.6)$$

To derive the differential equation implemented by this circuit, we are going to rewrite (1.6) in terms of the input and output voltage variables, i.e., V_1 and V_2, respectively. Toward that end, the capacitor current is written as

$$I_{cap} = C\frac{d}{dt}(V_2 - V_{D2})$$
$$= C\frac{d}{dt}V_2$$

(1.7)

because V_{D2} is a constant diode drop due to a fixed bias current I_o. It is given by

$$V_{D2} = V_T \cdot \ln\left(\frac{I_o}{I_S}\right)$$

(1.8)

where I_S and V_T are the diode saturation current and thermal voltage (more discussions on this diode relationship will be presented in Section 1.4.1).

On the other hand, the current through the diode D_1 can be written as

$$I_{D1} = I_S \cdot e^{V_{D1}/V_T}$$

(1.9)

Expressing the diode voltage V_{D1} with the terminal voltages V_1, V_2 and V_{D2}, (1.9) becomes

$$I_{D1} = I_S \cdot e^{\frac{[V_1 - (V_2 - V_{D2})]}{V_T}}$$
$$= I_S \cdot e^{\frac{V_1 - V_2}{V_T}} \cdot e^{\frac{V_{D2}}{V_T}}$$

(1.10)

Substitute (1.8) into (1.10), and simplify, we have

$$I_{D1} = I_o \cdot e^{\frac{V_1 - V_2}{V_T}}$$

(1.11)

At this point, we can apply the results in (1.7) and (1.11) to our KCL equation (1.6). We then arrive at

$$C\frac{d}{dt}V_2 = I_o \cdot e^{\frac{V_1 - V_2}{V_T}} - I_o$$

(1.12)

Multiply both sides by e^{V_2/V_T} and re-arranging, we have

$$I_o \cdot e^{\frac{V_2}{V_T}} + e^{\frac{V_2}{V_T}} \cdot C\frac{d}{dt}V_2 = I_o \cdot e^{\frac{V_1}{V_T}}$$

(1.13)

Apply the chain rule on the differentiator, and multiply both sides by I_S, we can write

$$I_S \cdot e^{\frac{V_2}{V_T}} + \frac{V_T C}{I_o} \cdot \frac{d}{dt}\left(I_S \cdot e^{\frac{V_2}{V_T}}\right) = I_S \cdot e^{\frac{V_1}{V_T}} \qquad (1.14)$$

which is the circuit equation describing a first-order filter section in the "log-domain". To mimic the linear first-order differential equation of (1.5), it is obvious that the filter input and output signals (assuming they are currents) should be related to V_1 and V_2 by

$$I_{in} = I_S \cdot e^{\frac{V_1}{V_T}} \qquad \text{and} \qquad I_{out} = I_S \cdot e^{\frac{V_2}{V_T}} \qquad (1.15)$$

Recognizing (1.15) is simply the *i-v* relationship of an ideal diode, it can be implemented by the circuits shown in Figure 1-4(c)-(d). They are the log and anti-log converters as stated before, by which the highly nonlinear (exponential) nature of the log-section is properly removed. As a result, (1.14) can be equivalently written as

$$I_{out} + \frac{V_T C}{I_o} \cdot I_{out}' = I_{in} \qquad (1.16)$$

which is of an identical form to (1.5) if the mapping of $I_{out} \Leftrightarrow V_o$ and $I_{in} \Leftrightarrow V_{in}$ are made. Therefore, when sandwiched between the log and anti-log converters, the diode-capacitor circuit is *exactly* implementing a linear first-order lowpass filtering function. The filter input and output are given by I_{in} and I_{out}, respectively.

Comparing the differential equations of the linear RC and the log-domain circuits, it is obvious that the R (resistance) term of (1.5) is now replaced by V_T/I_o in (1.16). Recall that V_T and I_o are the thermal voltage and the bias current of a diode. Therefore, a "diode in the log-domain" can be intuitively understood as "a resistor in the linear-domain". The advantage of (or, the motivation for) this diode-resistor is its electronic tunability: the resistance (and significantly, the resulting filter cutoff frequency) is now directly controlled by the current I_o.

1.2.2 Discussions

Although the solid-state technology 20 years ago was incomparable to what we have today, Adams has pointed out many key issues associated with log-domain filters in his 1979 paper. Many of the insights are still valid today[†]. Here, we will summarize several points:

- It was shown that how a highly nonlinear component, i.e., diode (which exhibits exponential characteristic), can be used to realize an overall linear system. This is achieved by embedding the log-circuit within the log and the anti-log converters.

- The cutoff frequency of the resulting log-domain filter, given by

$$f_{\text{cutoff}} = \frac{1}{2\pi} \cdot \left(\frac{I_o}{V_T C} \right) \tag{1.17}$$

  is controlled by the bias current I_o. It was pointed out that this equation is quite accurate over a wide range of currents (>60 dB). In other words, the resulting filter can exhibit decades of cutoff frequency tunability.

- According to (1.17), the filter corner is temperature-sensitive through the thermal voltage V_T, which equals kT/q. Therefore, f_{cutoff} will change in proportion to the absolute temperature in degrees Kelvin. Adams suggested that the effect could be greatly reduced by using complementary temperature compensation in the current source.

- As diodes demonstrate poor logarithmic characteristics, it was recommended that diode-connected (bipolar) transistors should be used.

- In one of his circuit examples, it was suggested that amplitude control could be accomplished by varying the bias current through an anti-log diode (not shown here). Therefore, beside cutoff frequency, the dc gain of the log-domain filters can also be conveniently and electronically controlled.

    Adams has moved on and applied the log-domain concept to several more complicated circuits. The Sallen-Key circuit shown in Figure 1-5(a) was one of his targets. In short, his philosophy on filter design (synthesis) can be stated as follows:

1. Replace each resistor in the linear prototype with a diode biased by a control current;

2. Add a level shifter to compensate for the diode drop for each diode introduced; and

3. Sandwich the resulting log-domain circuit between the log and anti-log converters to restore overall linearity.

This procedure results in the log-domain Sallen-Key filter displayed in Figure 1-5(b). However, when one attempts to derive the differential equation directly from the

---

†. In fact, even the circuit equations (1.6)-(1.16) (and the way they are derived) are quite similar to those of the log-domain filters to be discussed in later chapters. This reveals the fact that all these log-domain filters (old or new) are sharing the same fundamental concepts, although they may have very different appearances.

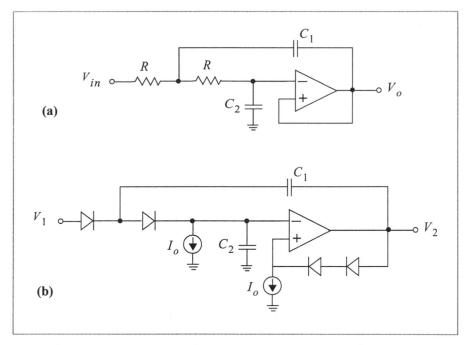

Figure 1-5: (a) Linear Sallen-Key filter, and (b) Adams's log- Sallen-Key filter (log and anti-log converters are not shown).

circuit, an error (non-linear) term will be observed. In other words, the design of the log-domain filter is not *exact*. The error term will cause the circuit to produce harmonic distortion even under ideal conditions. Although the above scheme works perfectly for the simple first-order *RC* circuit example (as discussed in the previous section), it is not generally applicable to other filter structures. The major drawback of his work was the lack of an error-free synthesis technique.

1.3   Some Recent Developments and Our Approach

The log-domain technique remained largely unexplored until 1993, when Frey proposed a general log-domain filter design approach known as "exponential-state-space synthesis" [22]. It involves transforming the variables in the state-space equations by certain exponential functions. As the mappings bear resemblance to the exponential I-V characteristic of a bipolar transistor, the resulting exponential state-space equation can be interpreted as a circuit equation with physical currents and voltages. As such, bipolar transistors and capacitors are used to realize the transformed state-space equation on a term by term basis. (More discussions on this synthesis method will be provided in Section 4.1.) Using this approach, Frey has successfully demonstrated the design of a biquadratic log-domain filter, and a seventh-order Chebyshev filter as a cascade of several biquads. Additional results based on these findings were later published in references [23]-[28]. Subsequently, a high-speed log-domain biquad formed entirely of npn transistors was developed and is reported in [17].

To simplify the filter synthesis and extend the approach to any order, Perry and Roberts developed the log-domain signal-flow-graph (SFG) approach [29]-[30]. Staying close to well-known filter design methods, the scheme resembles the method of operational simulation of *LC* ladder prototypes, a popular filter synthesis method used to realize high-order active-RC, MOS-C or g_m-C type integrated filter circuits. The result is that the log-domain filters maintain the low-sensitivity and low-noise properties of their *LC* prototypes. An important by-product of this approach is the insight into constructing linear systems from nonlinear elements with minimum linearization circuitry. This initial work was later confirmed experimentally through the development of several high-order, all-pole, lowpass log-domain filters. Later, the method was extended to include bandpass filters [31]. More recently, arbitrary filter functions, such as elliptic filters, were demonstrated through the use of a state-space formulation [32]-[33].

The intention of this book is to describe to the reader both the synthesis and analysis of high-order log-domain filter circuits. We shall make use of two different, but related, synthesis methods. The first is the method of operational simulation of *LC* ladder prototypes, followed by the lesser-known method based on a state-space formulation. Both methods attempt to capture the internal workings of the *LC* ladder in the active realization, thereby maintaining the low-sensitivity and low-noise properties of their *LC* prototypes. Subsequently, we shall extend the perturbation theory of *LC* ladder networks to the analysis of these log-domain filter circuits. In particular, we will explore how the filter nonidealities can be tackled in light of integrator magnitude and phase errors. We found this treatment intuitive and insightful, while at the same time, minimizing the need for complicated nonlinear mathematics.

1.4   Nonlinear Signal Processing in Log-Domain

For any practical system, an input to output linearity is always desired. However, it would be intriguing to enquiry: to achieve a *linear* system, must one always starts from *linear* building blocks? It is well known that transistors are nonlinear in nature: bipolar transistors are exponential, whereas MOS transistors are governed by a square-law. So far, tremendous efforts have been invested in linearizing these inherently nonlinear devices by elaborate circuit tricks, increasing power consumption, reducing operating speed and maintaining minute signal swing. Among them the technique of negative feedback is a good example. It is then natural to investigate if it is possible to utilize a transistor the way they intrinsically behave, while maintaining linear system operation. Maybe by doing so we can gain in terms of speed, distortion, power and circuit simplicity.

A natural starting point would be to review the Translinear Principle, which carries with it the connotation of "lying somewhere between the familiar home territories of the linear circuit and the formidable terrain's of the nonlinear" [34].

1.4.1 Translinear Principle

Translinear circuits achieve a wide range of algebraic functions by exploiting the current-to-transconductance relationship in bipolar transistors. Both the input and output signals to these circuits are in current form. In fact, the small voltage variations that result, which are typically less than a few tens of millivolts, are of incidental interest. The circuit function is essentially independent of the magnitude of the input signals, but instead depends on current ratios within the circuit. Desirably, the function is insensitive to temperature variations in the full range of operation on silicon. To illustrate the principle, we can begin with the fundamental expression relating the collector current, I_C, and the base emitter voltage, V_{BE}, as described by

$$I_C = I_S(T)e^{V_{BE}/V_T} \tag{1.18}$$

where V_T is the thermal voltage, kT/q, and $I_S(T)$ denotes the saturation current and its temperature dependence. It should be noted that I_S is a strong function of temperature: it can vary by 9.5% per °C [16]. When the device is driven by a certain V_{BE}, this level of temperature dependency will make the resulting I_C virtually unpredictable. As a result, one rarely sees bipolar devices driven in this manner in practice.

Conversely, when the transistor is driven by I_C to produce V_{BE}, the temperature dependence is now greatly reduced. Rewriting (1.18) as

$$V_{BE} = V_T \ln\left(\frac{I_C}{I_S(T)}\right) \tag{1.19}$$

an exact and linear relationship between the logarithm of I_C and V_{BE} is evident. When a couple of these devices are connected in a special manner to be demonstrated shortly, the resulting circuit can be made completely temperature independent. Furthermore, an impressive list of mathematical functions can be readily achieved. This leads us to the discussion of the Translinear Principle.

The principle of translinear circuits will be demonstrated on the general single loop network shown in Figure 1-6. In this closed loop configuration, we shall assume that all N devices are forward biased with some arbitrary voltage V_F. Also, we let N_1 represent the number of elements forward-biased in a clockwise (*CW*) direction, so that N_2 (= N- N_1) represents the total number of counter-clockwise (*CCW*) elements. To distinguish them, let the *CW* elements be assigned an even index. According to Kirchoff voltage law, the loop of junction voltages must sum up to zero,

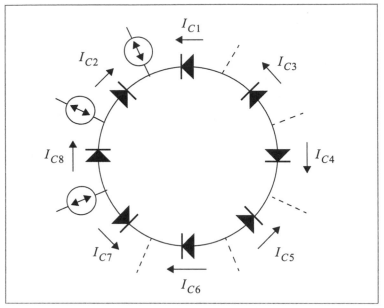

**Figure 1-6: A general closed-loop of forward-biased junction dem-
onstrating the Translinear Principle.**

$$\sum_{k=1}^{N_1} V_{F,2k} - \sum_{k=1}^{N_2} V_{F,2k-1} = 0 \qquad (1.20)$$

The junction voltage V_F will typically represent the base-emitter voltage V_{BE} of a bipolar device. By the same token, the junction current will correspond to the bipolar transistor collector current I_C. Therefore, based on (1.19), (1.20) can be rewritten as

$$\sum_{k=1}^{N_1} V_{T,2k} \ln\left(\frac{I_{C,2k}}{I_{S,2k}}\right) - \sum_{k=1}^{N_2} V_{T,2k-1} \ln\left(\frac{I_{C,2k-1}}{I_{S,2k-1}}\right) = 0 \qquad (1.21)$$

In a monolithic process where transistors are implemented in close proximity, it is generally valid to assume that the devices are operating at the same temperature, i.e., V_T's are all equal. Therefore, we can write

$$\sum_{k=1}^{N_1} \ln\left(\frac{I_{C,2k}}{I_{S,2k}}\right) - \sum_{k=1}^{N_2} \ln\left(\frac{I_{C,2k-1}}{I_{S,2k-1}}\right) = 0 \qquad (1.22)$$

Rearranging (1.22) results in

$$\prod_{k=1}^{N_1} \left(\frac{I_{C,2k}}{I_{S,2k}}\right) \cdot \prod_{k=1}^{N_2} \left(\frac{I_{S,2k-1}}{I_{C,2k-1}}\right) = 1 \tag{1.23}$$

To eliminate the dependency of (1.23) on temperature, the saturation current terms should cancel out. This would require that $N_1 = N_2$, and $N \ (= N_1 + N_2)$ be an even number. In other words, there must be equal number of CW and CCW elements connected together, and the loop must comprise an even number of elements. Therefore, we can write

$$\prod_{k=1}^{N/2} \frac{I_{S,2k}}{I_{S,2k-1}} = \lambda \tag{1.24}$$

where λ is a dimensionless number denoting their area ratio. Most often, when $\lambda = 1$, the areas of the bipolar transistors are identical, or they are well matched for pairs of oppositely connected elements. Equation (1.23) can then be rewritten as

$$\prod_{k=1}^{N/2} I_{C,2k} = \lambda \cdot \prod_{k=1}^{N/2} I_{C,2k-1} \tag{1.25}$$

This last equation captures the essence of the translinear principle developed by B. Gilbert. To summarize, it is re-stated as follows [16]:

"For any closed loop comprising any number of pairs of clockwise and counter clockwise forward-biased junctions, the product of currents for the elements in one direction is proportional to the corresponding product in the opposite direction. The factor of proportionality depends solely on the device geometry, and is essentially insensitive to process and temperature variations."

As an extension to the principle, when a voltage source V_s is introduced into the loop, (1.25) would become

$$\prod_{k=1}^{N/2} I_{C,2k} = \lambda \cdot e^{\frac{V_s}{V_T}} \cdot \prod_{k=1}^{N/2} I_{C,2k-1} \tag{1.26}$$

after a straightforward derivation which is left to the reader.

1.4.2 Translinear Circuit Examples

One of the earliest uses of the translinear principle was in realizing a wideband amplifier and an analog multiplier [35]-[36]. Here, we will describe the elegant example of a Type "B" two-quadrant translinear multiplier, as shown in Figure 1-7 [37]. Its operation will be described here to demonstrate the simplicity and

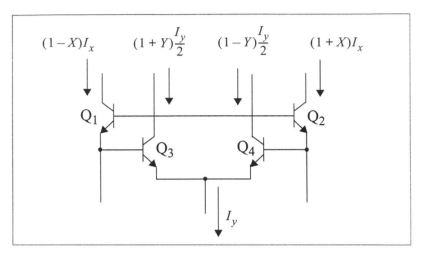

Figure 1-7: Type "B" two-quadrant translinear multiplier.

the practicality of the principle. The multiplier consists of four transistor arranged in a loop, two in each direction. Assuming the transistors are appropriately biased, collector currents of $(1 \pm X)I_x$ and $(1 \pm Y)I_y/2$ are generated in Q_1-Q_4, where X and Y are modulation indices lying between -1 to +1[†]. According to the translinear principle (1.25), we can write by inspection the following translinear relationship,

$$I_{C2} \cdot I_{C4} = I_{C1} \cdot I_{C3} \tag{1.27}$$

Now, if we substitute the appropriate transistor collector current as given in Figure 1-7, we can write

$$(1 + X)I_x(1 - Y)\frac{I_y}{2} = (1 - X)I_x(1 + Y)\frac{I_y}{2} \tag{1.28}$$

Equality occurs if and only if,

$$X = Y \tag{1.29}$$

If the output is taken as the differential current (I_z) between collectors of Q_3 and Q_4, the two-quadrant multiplication is achieved,

$$I_z = I_{C_3} - I_{C_4} = X \cdot I_y \tag{1.30}$$

[†]. The use of dimensionless modulation indices is often helpful in the analysis in translinear circuit, where the actual magnitudes of the currents are of secondary concern than their ratios.

where X represents the AC input signal, and the bias current I_y controls the multiplier gain. Superimposing two of these loops and sharing transistors Q_1 and Q_2 [36] achieves four-quadrant multiplication. It should be noted that the above analysis is an exact large-signal analysis, and is completely temperature insensitive. However, it does assume ideal translinear elements with perfect diode exponential property, zero ohmic resistances, infinite beta, and that they are perfectly matched.

Shown in Figure 1-8 is another example of the translinear principle. It is appropriately named the voltage-programmable current mirror [38]. Similarly, this circuit is composed of a single translinear loop with four complementary transistors. However, a voltage source, V_G, is now inserted into the loop. Taking this into account and according to the modified translinear principle in (1.26), we can directly write

$$I_{C2} \cdot I_{C4} = e^{\frac{V_G}{V_T}} \cdot I_{C1} \cdot I_{C3} \tag{1.31}$$

or with $I_{C1} = I_{C3} = I_{ref}$ and $I_{C2} = I_{C4} = I_{out}$ gives

$$I_{out} = I_{ref} \cdot e^{\frac{V_G}{2V_T}} \tag{1.32}$$

Here the output current I_{out} is expressed in terms of I_{ref} and the exponential difference of two voltages, $V_G = V_1 - V_2$. Obviously, a wide range of current gain is now realizable by altering V_G.

The second circuit just demonstrated happens to be the corner stone of the

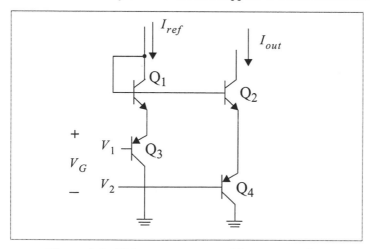

Figure 1-8: Voltage-programmable current mirror.

log-domain filtering technique. It is the key element for converting linear signals into their compressed form for signal processing, and expanding them to restore overall linearity. It will be re-introduced from a different perspective in the next section, and we will see how the translinear circuit can be applied in the frequency domain, producing compact and intriguing filter circuits.

1.4.3 *LOG* and *EXP* Operators

The circuit shown in Figure 1-8 whose behavior is described by (1.32) relates signals in a linear manner (such as I_{out}, I_{ref}) as well as those buried in the exponential function ($V_G = V_1 - V_2$). It is possible to employ this property to implement logarithmic signal compression, and likewise, exponential signal expansion[†]. Therefore, we will explicitly distinguish the compressed and uncompressed signals as log-domain and linear signals, respectively. These compression/expansion functions can be defined by the following pair of complementary mathematical operators[‡]:

$$LOG(x) = 2V_T \ln\left(\frac{I_o + x}{I_o}\right)$$

$$(1.33)$$

$$EXP(x) = I_o e^{\frac{x}{2V_T}} - I_o$$

Referring to the circuit shown in Figure 1-8 and (1.32), the logarithmic signal compression operator (*LOG*) can be implemented by:

1. setting I_{ref} to a linear input current $(1 + X) \cdot I_o$, where X is the modulation index in the range of $(0, 1)^*$;

2. forcing I_{out} to be the bias current I_o;

3. connecting V_1 to ground, and finally,

4. the logarithmically compressed (log-domain) signal will appear as

---

†. The concept of "logarithmic compression and exponential expansion" is actually identical to the "log and anti-log converters" as conceived by Adams. (See Section 1.2)

‡. Notice that these operators are slightly different from those presented by Perry and Roberts in [29] to more appropriately describe the log-domain integrator circuit in the next chapter.

*. It should be noted that physically, I_o is the bias current that carries the ac input current I_{in} on it. As common to all Class A circuits, the condition $|I_{in}| < I_o$ must be satisfied.

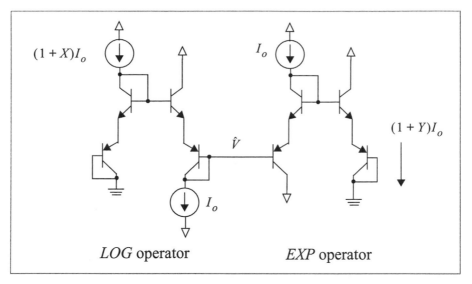

$(1 + X)I_o$

I_o

$(1 + Y)I_o$

$\hat{V}$

I_o

LOG operator EXP operator

Figure 1-9: Signal companding by *LOG* and *EXP* operators.

voltage V_2.

By the same token, the exponential signal expansion operator (*EXP*) is realized by:

1. applying the log-domain input voltage to V_1;

2. setting I_{ref} to be the bias current I_o;

3. connecting V_2 to ground, and finally,

4. the current I_{out} will be the exponentially expanded linear output current, which equals $(1 + Y) \cdot I_o$, where Y is the modulation index from 0 to 1.

The companding scheme discussed above is illustrated in Figure 1-9, in which *LOG* and *EXP* circuits are connected together. Voltage $\hat{V}$ is the log-domain (compressed) signal, while the linear signals are represented by XI_o and YI_o. Due to the inverse nature of *LOG* and *EXP* operators, i.e., $EXP(LOG(x)) = x$, X is identical to Y.

1.4.4 Linearization of Log-Domain System

A typical log-domain sub-circuit can be characterized by the SFG shown in Figure 1-10. The linear function $H(s)$, which can be summation, scaling, integration or any combination of them, is embedded between the *EXP* and *LOG* operators.

Signals $\hat{V}_i$ and $\hat{V}_o$ represent the log-domain inputs and output, respectively. Due to the *LOG*-linear-*EXP* format of the cell, an isolated log-domain transfer function, $\hat{V}_o / \hat{V}_i$, would be non-linear. In order to implement a practical linear system from this building block, linearization is undoubtedly necessary.

Suppose an arbitrary system as shown in Figure 1-11(a) is to be built using the log-domain circuit of Figure 1-10. Without loss of generality, $H_i(s)$ can be any linear mathematical function. One straight-forward way to tackle this problem (but rather redundant as will become obvious shortly) would be to abut external *LOG* and *EXP* blocks to the I/O of *each* log-domain sub-circuit. This will result in an overall linear input-output relationship as displayed in Figure 1-11(b).

A more economical way is to simply connect the log-domain sub-circuits together, and let the *LOG* and *EXP* operators cancel themselves naturally [29]. Note that this linearization takes place in both feedforward and feedback signal paths. The only extra components to add would be the input *LOG* and the output *EXP* blocks, as demonstrated in Figure 1-11(c)[†].

In summary, suppose we have a list of log-domain sub-circuits of the form shown in Figure 1-10 that implements a variety of functions. Then we would simply need to join these blocks together according to the specific topology, add the inverse

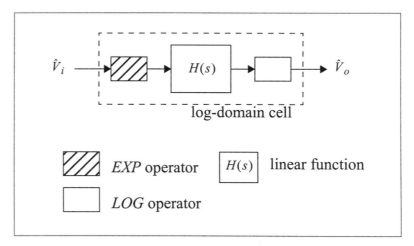

Figure 1-10: Signal-flow-graph of a typical log-domain sub-circuit.

---

†. As a remark, notice that the *LOG* and *EXP* of (1.33) are only of illustrative importance. In fact, the linearization and synthesis techniques discussed in this book will hold as long as the mathematical operators are *complementary*. Therefore, without loss of generality, there can be X-domain filters or systems, as long as the X operator and its complementary are easily realizable with practical circuits.

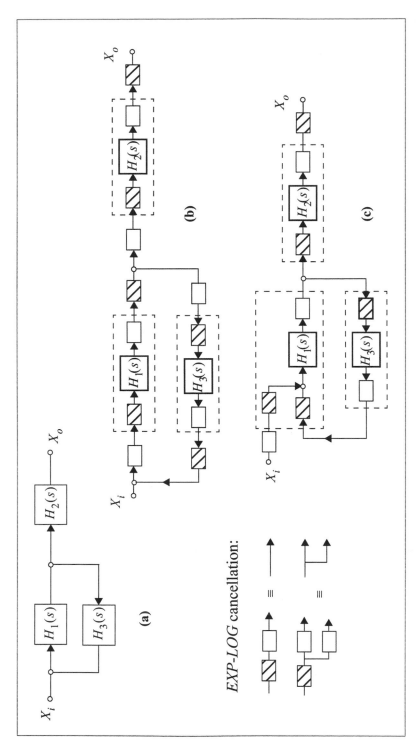

Figure 1-11: (a) SFG of an arbitrary system to be implemented in log-domain. **(b)** An implementation with individually linearized log-domain sub-circuits. **(c)** An economical log-domain system implementation.

operators at the input and output, and the desired linear system will result. As will be demonstrated in subsequent chapters, this linearization technique is the foundation of log-domain filter synthesis of this textbook, while the sub-circuits to be employed are known as log-domain integrators[†].

1.4.5 Potential Advantages of Log-Domain Filters

It is without doubt that our intention is to realize high-performance *linear* filters. Perhaps our plan to achieve this goal with highly *nonlinear* (log-domain) building blocks may sound totally absurd to some. Therefore, before we proceed any further, a few words on the potential advantages of the log-domain filter technique are in order.

Referring to Figure 1-11(c) and following the signal path, we can see that the linear input signal is logarithmic compressed (by the input *LOG*), processed by the subsequent log-domain circuits (i.e., the shaded blocks), and finally decompressed (by the output *EXP*). This intriguing signal handling feature, called companding, is indeed a well-known technique to improve signal integrity in transmission systems [20], [39]. Its effects of the achievable dynamic range can be qualitatively explained as follows.

Consider a practical analog system with certain noise and distortion properties. Assuming a sinusoidal signal is applied at the input. When its amplitude is very small, the signal may become indistinguishable from the noise generated by the circuit itself. To maintain certain signal-to-noise ratio, a lower bound on the signal amplitude is imposed. On the other hand, when the signal amplitude is increased gradually, one would observe a higher level of harmonic distortion generated by the circuit. Likewise, if a given signal linearity is to be ensured, an upper bound on the signal amplitude is required. The difference between the upper and the lower bounds of the signal can be understood as the useful signal range, or the dynamic range. This argument is graphically presented in Figure 1-12(a). Possible (and obvious) ways to enhance the dynamic range include lowering the noise floor (which frequently means bigger capacitors), and increasing the voltage supply. The tradeoffs are of course increases in silicon area and power consumption, which may be totally unacceptable under today's stringent cost and performance budget.

Signal companding offers another solution for dynamic range enhancement. The basic idea is to compress the signal to be processed, before noise and/or distortion of the processor has a chance to corrupt it. The scheme is illustrated in Figure 1-12(b). The input signal is first met by a compressor: the weak signal will be raised further above the noise floor, while the strong one will be shrunk away from the distortion-prone level. At the other end of the signal processor, a complementary

---

[†]. We will devote Chapter 2 to study the log-domain integrator (and its different variations) in detail. It will be demonstrated that the integrator indeed conforms to the paradigm *EXP-H(s)-LOG* as illustrated in Figure 1-10.

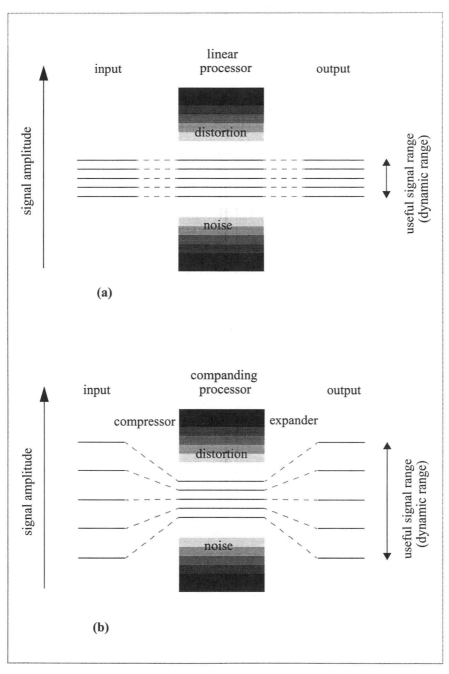

Figure 1-12: Comparison of dynamic ranges of (a) a conventional linear system, and (b) a companding system.

expander restores the signal's original dynamic range by attenuating the weak signal and amplifying the large signal. Signal is always kept away from the problematic zones at both upper and lower ends. Companding signal processor, such as our log-domain filter, can provide higher dynamic range than its linear counterpart.

For a similar reason, the log-domain filter is very suitable for very low power applications. The supply voltage can be kept to the minimum as the signal voltage is logarithmically compressed. Besides, unlike other traditional filter technologies, the integrators do not need to be linearized individually to achieve overall filter linearity. The exponential characteristics of the bipolar devices are directly employed, resulting in very simple integrator (and in turn, filter) circuits. Therefore, the related overhead in power consumption for linearizing the building blocks (say, the transconductors in g_m-C filters) can be avoided [44].

Due to companding, the voltage swing inside the log-domain filter is very small. The impedance level along the signal path is thus typically low. Charging and discharging of capacitors will be fast. These characteristics make the log-domain filters suitable for very high-speed applications[†]. Several experimental results seem to agree with this argument [17], [42].

1.5   An Integrator-Based Design and Analysis Approach

Adams constructed his log-domain filters by replacing the resistors in the linear prototype with diodes. For system linearity and biasing purposes, additional circuits such as current sources, level shifters, log and anti-log converters are added. It is fair to interpret the heart of this synthesis as: *the correspondence between "resistors in the linear-domain" and "diodes in the log-domain"*. Although this method works perfectly for a first-order filter, it gives rise to error terms for more complicated structures.

To remedy the problem, we are proposing a synthesis method based on: *the correspondence between "integrators in the linear-domain" and "integrators in the log-domain"*. In short, we are replacing each integrator (in the linear prototype) with its log-domain integrator equivalence. As such, our fundamental building blocks are shifted from "resistors" to "integrators". As discussed before, perfect linearization (cancellation of the logarithmic and exponential operators) can be achieved. The resulting log-domain filter can exactly mimic its linear equivalent (as will be verified

---

†. For a more thorough discussion of the potentials of companding processors, readers are encouraged to consult [20]. We have to admit that the application of companding at a circuit level is still in its research stage. There exist numerous uncertainties that are yet to be fully characterized and studied, such as the signal-dependency of noise, and the need of "coordinated" nonlinearities. Even the speed argument presented here can be questioned. High frequency performance may be limited because the compressed (distorted) waveform will contain harmonics which must be carefully preserved until expansion happens. Therefore, the useful signal bandwidth should always be limited to several times below the maximum operating speed of the circuit.

in later chapters).

In fact, most continuous-time filters reported are based on integrators not individual transistors or building blocks such as op amps [40]. This is probably due to the associated simplicity and modularity. Further advantages of this approach are stated below:

• It facilitates systematic filter synthesis. A filter can be viewed simply as an interconnection of integrators, each with different parameters such as unity-gain frequency and scale factor. Synthesis can then be done in a systematic and routine manner, regardless of the filter order. As a testimony, we will demonstrate in later chapters the design of high-order filters without any major increase in mathematical complexity.

• The research of log-domain filter design can then be tackled in two-fronts: (i) design of integrator circuit, and (ii) research on synthesis methods. This divide-and-conquer manner can help the log-domain filter field in general to mature at a faster pace. Circuit advancements can be brought to the log-domain integrators, resulting in high-speed, high-linearity, low-power log-domain filter realizations. Development in synthesis methods allows those building blocks to be put together systematically (or automatically, say, by CAD tools) for a versatile list of filter functions (such as Chebyshev, Elliptic, or even those with arbitrary filter shapes). Several intriguing integrator circuits will be presented and analyzed in Chapter 2, followed by the theories of log-domain filter synthesis in Chapters 3 and 4.

• Besides filter synthesis, the integrator-based study also significantly simplifies the task of analysis. Many well-proven theories exist that relate perturbations in the final filter response to underlying integrator errors. We will show that the analysis of log-domain filter nonidealities, albeit their intrinsic nonlinear signal processing nature, can indeed benefit substantially from the wealth of theories already developed for their linear equivalents. The log-domain filter nonideality analysis will be presented in Chapters 5 and 6.

1.6 Summary

With the advent of communication and signal-processing systems, there has been an increased emphasis on high-speed, high-linearity, low-power analog circuits. The log-domain filter is a novel form of continuous-time filter, which shows promise in these areas. As an introduction, we have presented the groundbreaking work by Adams. A first-order log-domain filter is described whose underlying concept and intriguing features are discussed. Unfortunately, the proposed design methodology will result in error terms when more complicated filter functions are attempted. A general and exact (error-free) synthesis method is necessary. This is the starting point of this text.

Seeing that a log-domain filter is essentially a linear system realized with intrinsically non-linear (exponential) bipolar devices, we reviewed the Translinear

Principle, which is a time-proven circuit technique that explicitly employs the diode characteristics for signal processing. We believe our log-domain research can benefit from the vast knowledge already developed in this area. In particular, it has come to our attention that the voltage-programmable current mirror seems to be a natural candidate to relate signals between the linear- and the log-domains. Based on this circuit, the complementary *LOG* and *EXP* mappings are introduced and a simple linearization scheme is derived which achieves complete cancellation of nonlinearities of the constituent log-domain building blocks. As a result, a perfectly linear log-domain filter can be achieved from highly nonlinear log-domain integrators. On the other hand, the filter is performing signal companding, which is an intriguing signal conditioning operation whose potential advantages were described.

CHAPTER 2   Log-Domain Integrators

As mentioned in the introduction, this book will follow an integrator-based approach for log-domain filter design and analysis. It is our belief that this would enhance general understanding about the subject, and avoid being quickly discouraged by the tedious mathematics. Based on a simple log-domain integrator circuit that involves only eight bipolar transistors and a capacitor, we will guide our readers through the circuit fundamentals in a step-by-step manner. Intriguing properties that are unique to the log-domain techniques will be highlighted. They will be followed by the discussions of several more sophisticated integrator designs. After finishing this chapter, the readers should have a solid understanding of the log-domain integration technique, and hopefully, appreciate the elegance and simplicity of these circuits, particularly in terms of their high-speed, low-power operation.

2.1 A Basic Log-domain Integrator

2.1.1 Detailed Analysis

Log-domain integrators are at the heart of log-domain filtering technique. To start, the voltage-programmable current mirror (Figure 1-8), together with its inverting counterpart, are re-drawn in Figure 2-1. In the log-domain literature, they are also called "log-domain cells". By writing KVL equations around the Q_1-Q_4 translinear loop, the basic log-domain equation is given by

$$I_{out} = I_o e^{\frac{\hat{V}_i - \hat{V}_o}{2V_T}} \tag{2.1}$$

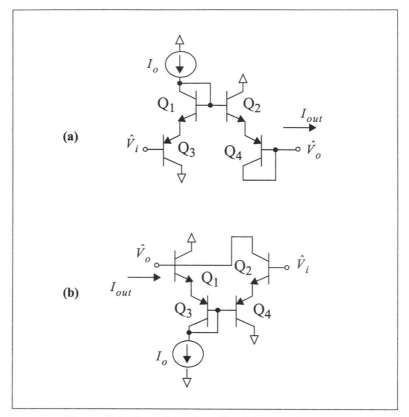

Figure 2-1: Log-domain cells with opposite polarities.

Combining the two log-domain cells of Figure 2-1, and adding a capacitor, the log-domain integrator [31] is formed as shown in Figure 2-2. Applying KCL at node P, we can write

$$C \cdot \frac{d\hat{V}_o}{dt} = I_o e^{\frac{\hat{V}_{ip} - \hat{V}_o}{2V_T}} - I_o e^{\frac{\hat{V}_{in} - \hat{V}_o}{2V_T}} \qquad (2.2)$$

where $\hat{V}_{ip}$, $\hat{V}_{in}$ and $\hat{V}_o$ denote the log-domain positive input, negative input, and log-domain output, respectively. Multiplying through by $\exp(\hat{V}_o/(2V_T))$ and applying the chain rule will result in

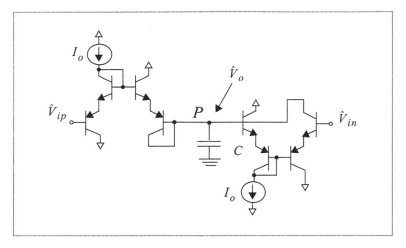

Figure 2-2: Log-domain positive and negative integrator pair.

$$\frac{2V_T}{I_o} \cdot C \cdot \frac{d}{dt}\left\{ I_o e^{\frac{\hat{V}_o}{2V_T}} - I_o \right\} = \left\{ I_o e^{\frac{\hat{V}_{ip}}{2V_T}} - I_o \right\} - \left\{ I_o e^{\frac{\hat{V}_{in}}{2V_T}} - I_o \right\} \qquad (2.3)$$

If we define a pair of inverse *LOG* and *EXP* mappings as in (1.33), which is recaptured below,

$$LOG(x) = 2V_T \ln\left(\frac{I_o + x}{I_o}\right) \qquad EXP(x) = I_o e^{\frac{x}{2V_T}} - I_o, \qquad (2.4)$$

we can rewrite (2.3) as

$$EXP(\hat{V}_o) = \frac{I_o}{2V_T} \cdot \frac{1}{C} \cdot \int \{EXP(\hat{V}_{ip}) - EXP(\hat{V}_{in})\} dt \qquad (2.5)$$

This can also be symbolically represented by the SFG shown in Figure 2-3. Notice that the integrator indeed conforms to the paradigm shown in Figure 1-10 presented in the previous chapter. Therefore, the linearization scheme (by natural cancellation) will undoubtedly apply to these circuits.

Having gone through the above routine mathematical manipulations, we would like to point out a few interesting features.

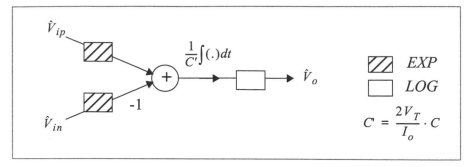

Figure 2-3: Log-domain integrator Signal-Flow-Graph.

2.1.2　Highlights of Intriguing Features

As described in the previous chapter, Adams first invented log-domain filters for easy tunability. As revealed from (2.5), the bias current I_o can be viewed as to "scale" the capacitor. It is this factor that accounts for the electronic tunability of this integrator and log-domain filters[†] in general.

Damped integration in the log-domain can be achieved by applying the integrator output signal, $\hat{V}_o$ to the negative input, $\hat{V}_{in}$. The circuit equation is given by substituting $\hat{V}_{in} = \hat{V}_o$ into (2.2), which becomes

$$C \cdot \frac{d\hat{V}_o}{dt} = I_o e^{\frac{\hat{V}_{ip}-\hat{V}_o}{2V_T}} - I_o \tag{2.6}$$

The second term on the right hand side indicates a dc current being drawn away from the capacitor node. Therefore, log-domain damped integrator can be realized by replacing the right log-domain cell of Figure 2-2 by a dc current source [29]. The damped log-domain integrator is shown in Figure 2-4. As will be appreciated in Chapters 5 and 6, electronic tunability and damping are the keys to understanding log-

---

† . The factor $1/V_T$ also makes the integrator and the resulting filter temperature-dependent. According to the design method described in this book [29], any fluctuations in operating temperature (T, in °C) from 25°C will introduce the following scalar integrator error k,

$$k = \frac{273.15 + 25}{273.15 + T}$$

As will become evident later, k will also represent the final filter cutoff/center frequency deviation. For a temperature insensitive design, PTAT current bias is therefore necessary.

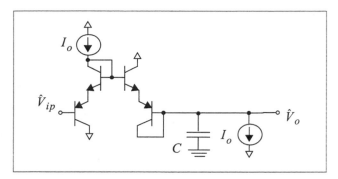

Figure 2-4: Damped log-domain integrator.

domain filter nonidealities and to compensate for their effects.

2.2 Input and Output Stages

We have seen from Chapter 1 that in order to maintain the overall linearity of a log-domain system, a *LOG* block must be added to the input, while an *EXP* block is required at the output. This procedure will use to linearize the first-order log-domain filter shown in Figure 2-5(a). An input *LOG* cell converts the linear input current into a log-domain voltage. The damped log-domain integrator then performs the filtering function. The resulting log-domain output voltage is then expanded by the *EXP*-cell to restore the linear output current. The reader will find that the results shown here are applicable to all of the circuits shown in this book.

Replacing the different blocks of Figure 2-5(a) by their appropriate circuit equivalence, Figure 2-5(b) is drawn. Experienced circuit designers will probably suspect that there is a certain amount of redundancy between the *LOG* cell and the integrator. (Notice that the two transistors on both sides of signal $\hat{V}_{input}$ are implementing nothing more than level shift. Intuitively, they could be eliminated.) Let us perform some algebraic manipulations to implement a more efficient circuit. For the *LOG* circuit, the linear input current I_{input} is converted into a log-domain voltage $\hat{V}_{input}$,

$$\hat{V}_{input} = 2V_T \log\left(\frac{I_o + I_{input}}{I_o}\right) \tag{2.7}$$

This voltage is in turn applied to the input of the damped log-domain integrator. So, we can substitute $\hat{V}_{input}$ of (2.7) into $\hat{V}_{ip}$ of (2.6) and with the appropriate change of variables, i.e., $\hat{V}_o \rightarrow \hat{V}_{output}$. After trivial simplification, we have

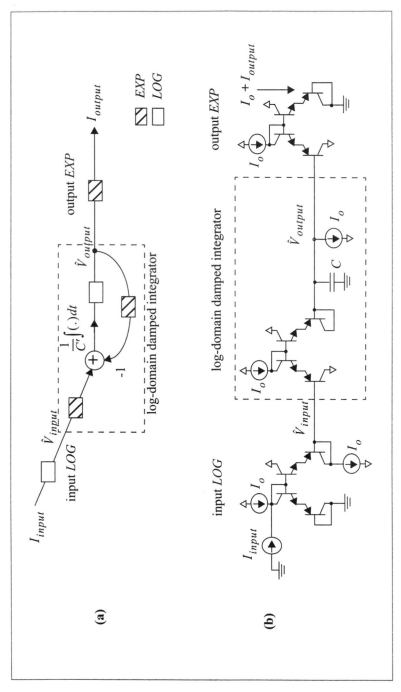

Figure 2-5: Input *LOG* and output *EXP* demonstration: (a) a log-domain system example, (b) straight-forward circuit implementation, and (c) an efficient realization (*to be continued on next page*).

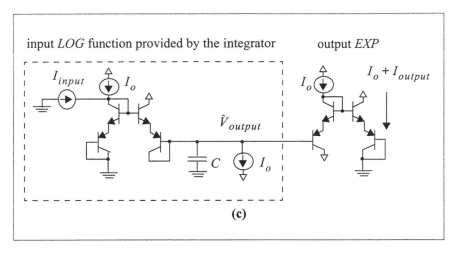

Figure 2-5: (continued) Input *LOG* and output *EXP* demonstration.

$$C \cdot \frac{d\hat{V}_{output}}{dt} = (I_o + I_{input}) \cdot e^{\frac{-\hat{V}_{output}}{2V_T}} - I_o \tag{2.8}$$

The first term of the right hand side denotes a log-domain cell whose current source (I_o) is summed with the input current (I_{input}), while the positive voltage input is grounded ($\hat{V}_{ip} = 0$). Therefore, (2.8) can be efficiently implemented as in Figure 2-5(c). In other words, the damped log-domain integrator can intrinsically accomplish the input *LOG* function. This result can equally be applied to any non-damped log-domain integrator. In fact, in order to follow the systematic synthesis procedure of this text, the input current will always be associated with an integrator cell whose input voltages are both grounded as shown in Figure 2-6.

2.3   Further Log-Domain Integrator Examples

At this point in our discussion it might be a good place to investigate a few types of log-domain integrator structures. Our list of circuits described here is by no means exhaustive. Our intention here is merely to articulate the trend in which these circuits evolve, particularly in the direction of low-power and high-speed applications. Nevertheless, we should see that all of them share very similar formulations. In fact, it is crucial to observe that any circuits that implement some forms of equation (2.2) will qualify to serve as log-domain integrators. For consistency, each of them will be described using similar formulations as in the previous section. They can all be employed as the building blocks for the log-domain filter synthesis described later in this book, regardless of the fact that they may not share the identical *LOG/EXP* operators as (2.4).

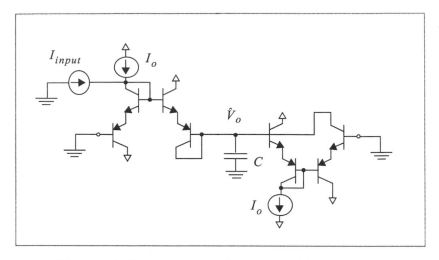

Figure 2-6: Input stage to an integrator with capacitor C.

2.3.1 High-Speed Log-Domain Integrators

Unless a costly bipolar process featuring both fast npn and pnp devices is available, the speed of the log-domain filters built using complementary devices will always be severely limited by the poor high-frequency performance of the pnp transistor. An all-npn log-domain integrator is needed to fully utilize the high speed potential a modern technology can offer. Figure 2-7 shows one example. The subcircuit Q_1-Q_2 on the left and its biasing is a standard cell seen in many log-domain filters [17]-[19], [41]. They form the first two-transistor translinear loop. The remaining circuitry can be recognized as the negative exponential transconductance cell reported in [17]. Transistors Q_3 and Q_4 form another translinear loop, while Q_5, Q_6 and the resistor are only included to satisfy biasing requirement[†]. The two transistor pairs: Q_1-Q_2 and Q_3-Q_4 respectively perform the functions of positive and negative log-domain cells similar to the circuits of Figure 2-1. Two currents are produced at the collectors of Q_2 and Q_4:

$$I_{C2} = I_o e^{\frac{\hat{V}_{ip} - \hat{V}_o}{V_T}} \quad ; \quad I_{C4} = I_o e^{\frac{\hat{V}_{in} - \hat{V}_o}{V_T}} \tag{2.9}$$

Notice the similarity between these expressions and (2.1). To realize the log-domain integration, the difference of these currents will be pumped into an integrating

---

[†]. Transistors Q_5, Q_6 and the resistor are designed such that the bias of Q_6 will be flexible enough to take any current from the translinear loop Q_3-Q_4. Notice that this current is not fixed, but a signal-dependent quantity.

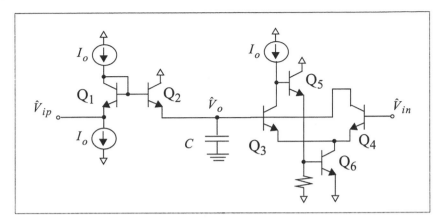

Figure 2-7: Log-domain integrator with all npn transistors.

capacitor. Routinely writing a KCL equation at the capacitor node, we have

$$C \cdot \frac{d\hat{V}_o}{dt} = I_o e^{\frac{\hat{V}_{ip} - \hat{V}_o}{V_T}} - I_o e^{\frac{\hat{V}_{in} - \hat{V}_o}{V_T}} \qquad (2.10)$$

Clearly, this equation is almost identical to (2.2) except for a scalar difference in the arguments of the exponential operations. Following the mathematical manipulations similar to those employed to derive (2.2) to (2.5), and defining a new pair of complementary *LOG/EXP* mappings as

$$LOG(x) = V_T \ln\left(\frac{I_o + x}{I_o}\right) \qquad EXP(x) = I_o e^{\frac{x}{V_T}} - I_o, \qquad (2.11)$$

the following log-domain integration equation will result,

$$EXP(\hat{V}_o) = \frac{I_o}{V_T} \cdot \frac{1}{C} \cdot \int \{EXP(\hat{V}_{ip}) - EXP(\hat{V}_{in})\} dt \qquad (2.12)$$

Revealed by the similarity of their formulations, this circuit will share all the characteristics of the log-domain integrator discussed before. Since the entire signal path is composed *only* of npn transistors, this integrator circuit can operate up to hundred's of MHz [17]-[18], which is suitable for very high frequency applications.

It is well known that differential implementations exhibit better common-mode noise rejection and higher linearity than their single-ended counterparts. Any substrate noise, such as those induced from the digital circuits in a mixed-signal environment, will be injected equally into two symmetrical circuits and have no effects. Besides, even-order distortions due to circuit nonlinearities will also be mutually cancelled at the output. A fully differential log-domain integrator can be simply obtained by using two single-ended integrators and cross-coupling their inputs as shown in Figure 2-8.

By inspection, integration expressions similar to (2.12) can be written for the positive and negative log-domain output voltages, $\hat{V}_{op}$ and $\hat{V}_{on}$, as

$$
\begin{aligned}
EXP(\hat{V}_{op}) &= \frac{I_o}{V_T} \cdot \frac{1}{C} \cdot \int \{EXP(\hat{V}_{ip}) - EXP(\hat{V}_{in})\} dt \\
EXP(\hat{V}_{on}) &= \frac{I_o}{V_T} \cdot \frac{1}{C} \cdot \int \{EXP(\hat{V}_{in}) - EXP(\hat{V}_{ip})\} dt
\end{aligned}
\tag{2.13}
$$

In terms of a differential log-domain output voltage, $\hat{V}_o$, which relates to $\hat{V}_{op}$ and $\hat{V}_{on}$ by $EXP(\hat{V}_o) \equiv EXP(\hat{V}_{op}) - EXP(\hat{V}_{on})$, (2.13) can be rewritten as

$$
EXP(\hat{V}_o) = \frac{I_o}{V_T} \cdot \frac{2}{C} \cdot \int \{EXP(\hat{V}_{ip}) - EXP(\hat{V}_{in})\} dt
\tag{2.14}
$$

Comparing (2.14) and (2.12), it can be seen that the differential structure has the

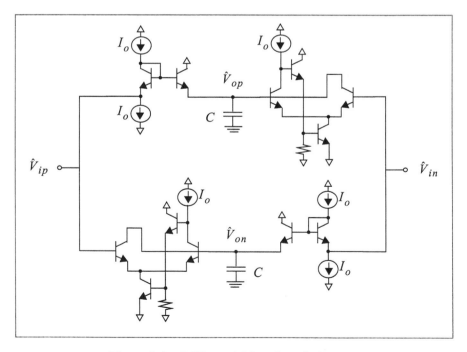

Figure 2-8: Differential log-domain integrator.

effect of doubling the sizes of the integrating capacitors needed to achieve the same center/cutoff frequency than by using a single-ended structure. Although it means an increase in consumption of capacitor area, it also helps to avoid the integrating capacitor from becoming too small and becoming comparable to the various parasitic capacitances. This is desirable at VHF to ensure reasonable predictability of the frequency response and Q-factor while keeping a low current biasing level [42].

A careful look at the circuit of Figure 2-8 reveals that the quiescent voltages at the output terminals (i.e., $\hat{V}_{op}$ and $\hat{V}_{on}$) of the integrator are expected to be sensitive to nonidealities and operating conditions such as device mismatch, thermal drift, etc. As described by (2.10), any slight variation of these voltages will severely affect the dc current biasing as they are exponentially related. As a result, the filter will exhibit rather unpredictable transfer characteristics. In order to keep the quiescent output voltages close to their desired values; a common-mode feedback (CMFB) circuit can be used. We refer the reader to [42] for the details regarding the CMFB implementation.

Examining the integrator circuit carefully, it can be observed that no node is more than 3 diode drops away from the power rails. Therefore, the minimum supply voltage can be lowered to as low as 3 V [17]. This is an encouraging feature in the sense that it meets the typical voltage budget for modern analog electronics. However, we can also argue that to accommodate the (compressed) log-domain signals which are typically limited to a few tens of mV's (i.e., several V_T's), a supply voltage of 3 V seems excessive. In fact, this supply voltage is largely dictated by the biasing circuitry, which commonly involves the stacking of transistors. Therefore, it can be contemplated that by means of smart circuit tricks that eliminate the need for cascoding, there is still a lot of room for supply voltage reduction. This will be explored further in the next section.

2.3.2 Low-Power Log-Domain Integrators

To explore the low power potential of log-domain techniques, a BiCMOS integrator circuit of Figure 2-9 is derived [43]-[44]. Bipolar npn transistors are used to form the translinear loop necessary to perform signal companding and integration, while all supporting circuits and current mirroring are realized by MOS transistors. MOSFETs are selected because they have the advantages of zero gate current (infinite DC current gain) and lower saturation voltages than their bipolar counterparts. The obvious drawback is that the maximum operating speed of the integrator is now limited by the pmos transistors.

The log-domain integrator of Figure 2-9 is a very compact and elegant design, in which the positive and negative log-domain cells are tightly coupled together. The diode-connected transistor Q_1 provides the circuit path for positive integration, while Q_2, M_1-M_2 realize the negative one. Transistors Q_3 and M_3 implement a level shifter that is also responsible for the frequency tuning capability. To demonstrate the operation of this integrator, we will first derive the collector

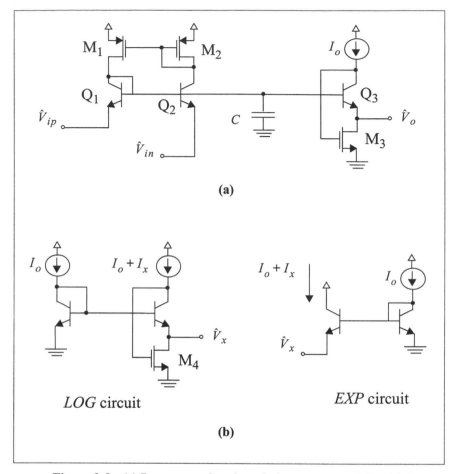

Figure 2-9: (a) Low-power log-domain integrator, and (b) *LOG*
and *EXP* circuits.

current of Q_1 by writing a KVL equation across the translinear loop consisting of Q_1
and Q_3,

$$\hat{V}_{ip} + V_T \ln\left(\frac{I_{C1}}{I_S}\right) - V_T \ln\left(\frac{I_o}{I_S}\right) = \hat{V}_o \qquad (2.15)$$

Rearrange, we get

$$I_{C1} = I_o e^{\frac{\hat{V}_o - \hat{V}_{ip}}{V_T}} \qquad (2.16)$$

Similarly, an equation can be written for the collector current of Q_2 by writing a KVL equation around transistor Q_2 and Q_3,

$$I_{C2} = I_o e^{\frac{\hat{V}_o - \hat{V}_{in}}{V_T}} \tag{2.17}$$

It is easy to recognize that I_{C1} and I_{C2} are remarkably similar to the output currents of the log-domain cell, or (2.1). To achieve the integrator function such as (2.2), we need circuits that find the difference between them and pump it into an integrating capacitor. The current mirror M_1-M_2 provides this function. The capacitor current is found by writing a KCL equation at the capacitor node,

$$I_C = I_{C2} - I_{C1} \tag{2.18}$$

which is also related to $\hat{V}_o$ by

$$
\begin{aligned}
I_C &= C \cdot \frac{d}{dt}\left(\hat{V}_o + V_T \ln\left(\frac{I_o}{I_S}\right) \right) \\
&= C \cdot \frac{d}{dt} \hat{V}_o
\end{aligned}
\tag{2.19}
$$

Substituting the expressions of (2.16), (2.17) and (2.19) into (2.18), we arrive at the familiar log-domain integrator equation,

$$C \cdot \frac{d}{dt} \hat{V}_o = I_o e^{\frac{\hat{V}_o - \hat{V}_{in}}{V_T}} - I_o e^{\frac{\hat{V}_o - \hat{V}_{ip}}{V_T}} \tag{2.20}$$

which has a very similar form to that of the basic log-domain integrator function, (2.2). Routinely applying the mathematical manipulations as in (2.3), we have

$$\frac{V_T}{I_o} \cdot C \cdot \frac{d}{dt}\left\{ I_o e^{\frac{-\hat{V}_o}{V_T}} - I_o \right\} = \left\{ I_o e^{\frac{-\hat{V}_{ip}}{V_T}} - I_o \right\} - \left\{ I_o e^{\frac{-\hat{V}_{in}}{V_T}} - I_o \right\} \tag{2.21}$$

We can then define a new pair of complementary *LOG* and *EXP* operators as

$$LOG(x) = -V_T \ln\left(\frac{I_o + x}{I_o}\right) \qquad EXP(x) = I_o e^{\frac{-x}{V_T}} - I_o \tag{2.22}$$

which the circuits shown in Figure 2-9(b) can implement. The simple verification step is left to the readers. Finally, we can rewrite (2.21) as

$$EXP(\hat{V}_o) = \frac{I_o}{V_T} \cdot \frac{1}{C} \cdot \int \{EXP(\hat{V}_{ip}) - EXP(\hat{V}_{in})\} dt \qquad (2.23)$$

This equation describes the standard log-domain integrator behavior whose SFG is identical to that given in Figure 2-3 (except that $C' = (V_T/I_o)C$). The bias current in Q_3 is responsible for the integrator center frequency tuning. Notice that the minimum supply voltage of this integrator is given by the sum of the source-to-gate voltage (V_{sg}) of M_2 and the saturation voltages (V_{sat}) of Q_2 and M_4 (of the input LOG cell). With typical V_{sg} and V_{sat} equal 0.6 V and 0.2 V respectively, the minimum supply voltage can be decreased to as low as 1.0 V. This is a significant reduction to the supply voltage of 3 V as required by the high-speed integrator discussed in Section 2.3.1.

The above integrator can be extended to a fully differential structure as shown in Figure 2-10 [45]. Since both positive and negative integrator outputs exist, current mirrors (such as M_1 and M_2 of Figure 2-9) are no longer needed for signal inversion. The positive and negative signals will now see an identical path, and the circuit is perfectly symmetrical. This is a property not enjoyed by the single-ended circuit.

The operation of the differential integrator is described below. Two currents similar to (2.16) and (2.17) are generated by the translinear loops formed by Q_1-Q_2, Q_4-Q_5 respectively,

$$I_{C1} = I_o e^{\frac{\hat{V}_{op} - \hat{V}_{ip}}{V_T}} \quad ; \quad I_{C4} = I_o e^{\frac{\hat{V}_{on} - \hat{V}_{in}}{V_T}} \qquad (2.24)$$

They are then combined with the capacitor currents, I_{cp} and I_{cn}, and mirrored by M_3-M_4, and M_5-M_6. As such, the collector currents of Q_6 and Q_3 is given by

$$I_{C6} = I_{C1} + I_{cp} \quad ; \quad I_{C3} = I_{C4} + I_{cn} \qquad (2.25)$$

If we further express these currents in terms of their corresponding log-domain voltages, (2.25) can be rewritten as

$$I_{C6} = I_o e^{\frac{-\hat{V}_{on}}{V_T}} = I_o e^{\frac{\hat{V}_{op} - \hat{V}_{ip}}{V_T}} + C\frac{d}{dt}\hat{V}_{op} \qquad (2.26)$$

and

$$I_{C3} = I_o e^{\frac{-\hat{V}_{op}}{V_T}} = I_o e^{\frac{\hat{V}_{on} - \hat{V}_{in}}{V_T}} + C\frac{d}{dt}\hat{V}_{on} \qquad (2.27)$$

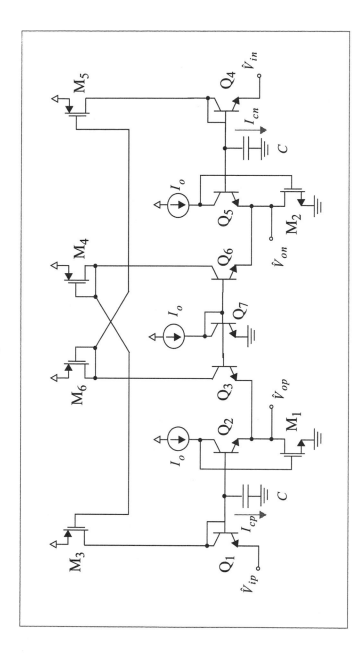

Figure 2-10: Differential low-power log-domain integrator.

Then, by routine mathematical manipulations, which the readers should be familiar with by now, (2.26) and (2.27) can be written as

$$\left[I_o e^{\frac{-\hat{V}_{ip}}{V_T}} - I_o\right] + \frac{CV_T}{I_o} \cdot \frac{d}{dt}\left[I_o e^{\frac{-\hat{V}_{op}}{V_T}} - I_o\right] = I_o e^{\frac{-\hat{V}_{on} - \hat{V}_{op}}{V_T}} \qquad (2.28)$$

and

$$\left[I_o e^{\frac{-\hat{V}_{in}}{V_T}} - I_o\right] + \frac{CV_T}{I_o} \cdot \frac{d}{dt}\left[I_o e^{\frac{-\hat{V}_{on}}{V_T}} - I_o\right] = I_o e^{\frac{-\hat{V}_{on} - \hat{V}_{op}}{V_T}} \qquad (2.29)$$

Applying the same mapping as (2.22), and subtracting (2.29) from (2.28), we arrive at

$$EXP(\hat{V}_{ip}) - EXP(\hat{V}_{in}) + \frac{CV_T}{I_o} \cdot \frac{d}{dt}[EXP(\hat{V}_{op}) - EXP(\hat{V}_{on})] = 0 \qquad (2.30)$$

Defining the differential log-domain output as $EXP(\hat{V}_o) \equiv EXP(\hat{V}_{op}) - EXP(\hat{V}_{on})$, and rearranging (2.30), the derivation of the log-domain integrator function is complete, as we obtain the familiar expression (albeit, with a minus sign),

$$EXP(\hat{V}_o) = -\frac{I_o}{V_T} \cdot \frac{1}{C} \cdot \int[EXP(\hat{V}_{ip}) - EXP(\hat{V}_{in})]dt \qquad (2.31)$$

Class AB operation allows the signal currents to be much greater than the quiescent current, thus resulting in very efficient and low power consumption. Similar to traditional differential systems, the signals in log-domain filters will be represented by the difference of two physical currents or voltages. It was suggested in [45] that an input conditioner could be used to perform the signal splitting and *LOG* compression at the input of the log-domain system. The input signal is represented by the difference of two strictly positive currents (I_{inp} and I_{inn}), whose common-mode component is determined by the geometric mean law,

$$I_{inp} \cdot I_{inn} = I_Q^2 \qquad (2.32)$$

where the constant I_Q can be physically interpreted as the bias current. For instance, to represent a large positive signal, it is expected that I_{inn} will become very small while I_{inp} approaches infinity. A *LOG* operator then converts them into log-domain voltages, which are in turn applied to different parts of the log-domain filter. These voltages will be centered on a dc voltage and follow an arithmetic mean law[†]. Readers can refer to [43] for the circuit implementation.

2.3.3 High-Speed Low-Power Log-Domain Integrators

For high-speed log-domain integrator design, an all npn realization is required. The circuit shown in Figure 2-7 of Section 2.3.1 fulfils this goal. However, the stacking of transistors sets the lower limit of supply voltage unnecessarily high. A low power version is proposed in Figure 2-9(a), in which the cascoding stages are largely eliminated. Unfortunately, this requires the use of p-type transistors (pmos or pnp), which undermine its maximum achievable speed. At this point, a natural question to ask is do we have a log-domain integrator that is *both* fast and low power? That is, an integrator circuit that involves only npn transistors that do not stack up on one another?

Figure 2-11 shows a very simple circuit [46] that tries to answer the above question. It is composed out of two identical log-domain cells, Q_1-Q_2, and Q_5-Q_6. Notice these translinear building blocks are present in almost all integrator circuits described before. It is the translinear nature of the circuit that generates the log-domain currents. A voltage buffer realized by Q_5-Q_6 and their biasing (enclosed within the dashed lines) is inserted between them. Therefore, the output log-domain voltage $\hat{V}_o$ at the emitter of Q_2 also appears at the emitter of Q_4. Therefore, we can write the log-domain currents as,

$$I_{C2} = I_o e^{\frac{\hat{V}_{ip} - \hat{V}_o}{V_T}} \quad ; \quad I_{C4} = I_o e^{\frac{\hat{V}_{in} - \hat{V}_o}{V_T}} \tag{2.33}$$

In order to achieve a log-domain integrator equation, we need to realize the difference of these two currents on a capacitor node. Therefore, current mirror Q_7-Q_8 is employed. It is clear that Q_7 will carry a current that is equal to I_{C4}. Mirrored by Q_8, this current will be pulled away from the capacitor node. In order words, this current mirror effectively performs signal inversion for the log-domain cell on the right. The KCL equation on the capacitor node becomes

---

†. To show this property, we can write

$$\hat{V}_{inp} + \hat{V}_{inn} = -V_T \ln\left(\frac{I_{inp}}{I_o}\right) - V_T \ln\left(\frac{I_{inn}}{I_o}\right)$$

$$= -V_T \ln\left(\frac{I_{inp} \cdot I_{inn}}{I_o^2}\right)$$

$$= -2V_T \ln\left(\frac{I_Q}{I_o}\right)$$

Notice that if we set $I_Q = I_o$, the differential log-domain voltages will center around 0 V.

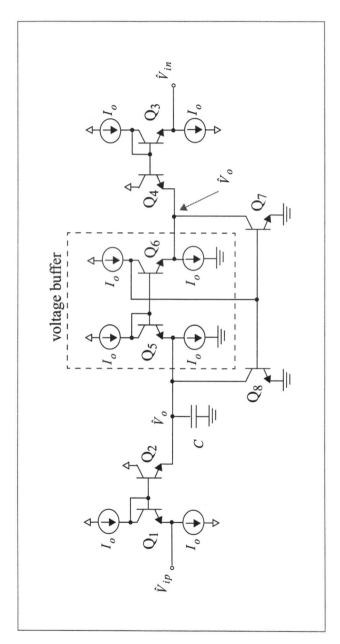

Figure 2-11: High-speed low-power log-domain integrator.

$$C \cdot \frac{d\hat{V}_o}{dt} = I_o e^{\frac{\hat{V}_{ip} - \hat{V}_o}{V_T}} - I_o e^{\frac{\hat{V}_{in} - \hat{V}_o}{V_T}} \qquad (2.34)$$

which is identical to (2.10). Therefore, this circuit will realize the identical log-domain integrator function as that of Figure 2-7 as

$$EXP(\hat{V}_o) = \frac{I_o}{V_T} \cdot \frac{1}{C} \cdot \int \{ EXP(\hat{V}_{ip}) - EXP(\hat{V}_{in}) \} dt \qquad (2.35)$$

Notice that the input and output voltages of the integrator in Figure 2-11 are at the same dc level. Therefore, filter synthesis can be easily achieved by directly coupling these integrators. The minimum supply voltage necessary for the circuit is one diode drop (V_{BE}) plus the saturation voltages (V_{sat}) of the positive current sources. Typically speaking, for $V_{BE} = 0.6V$ and $V_{CE} = 0.2V$, it is possible to keep the supply voltage as low as 1 V. Since the entire circuit is built using npn transistors, it is suitable for high-speed applications. It is demonstrated in [46] that for a cheap bipolar process featuring npn transistors with $f_T = 2.5$ GHz and off-chip capacitors of 1 nF, the log-domain filter can operate up to 2 MHz under a 1.2 V supply. Moving the capacitors on chip and reducing their values by several orders of magnitude, the maximum operating frequency range should reach the 100's MHz range. This has yet to be demonstrated.

2.4 Summary

The fundamental concept of a log-domain integrator is described, while its input-output behavior was quantified. A pair of log-domain cells with opposite polarities and an integrating capacitor form the core of the integrator. Further, the integrator time constant can be electronically controlled by its bias current. This factor accounts for the simple tunability of log-domain filters. On the other hand, a damped integrator in the log-domain can be realized by simply replacing one log-domain cell by a dc current source.

From the basic low-speed high-power log-domain integrator of Figure 2-2 to the high-speed low-power version of Figure 2-11, we have described the recent circuit developments in log-domain integrators. In a nutshell, the relatively slow pnp transistors incorporated in the early cells were eliminated to allow for high-speed operations. As well, for low power applications, transistor stacks were eliminated as much as possible so that the supply voltage can be lowered.

Although these circuits may have very different appearances, they in fact behave in similar ways. The readers can probably realize that the equations (and their derivations) for the basic structure, such as (2.1) to (2.5) of Section 2.1.1, keep repeating themselves for the integrator circuits in Sections 2.3.1-2.3.3. Most importantly, we could see that any circuit that implements some forms of (2.2) could be employed as a log-domain integrator [31].

Finally, Table 2.1 summarizes the results of several log-domain filter experiments featuring different log-domain integrators. They are categorized similar to the organization of this chapter: (i) basic design, (ii) high-speed (HS) design, (iii) low-power (LP) design, and (iv) high-speed low-power (HSLP) design. These results should convey a quick overview for the achievable performances of the various integrator circuits. However, a fair direct comparison may not be feasible due to the very different technologies, filter orders, types (bandpass/lowpass), and the form of signal (Class A/AB, differential/single-ended) in question. Even the experimental setup, such as wafer probing versus conventional chip testing in package, may have crucial influences on the measurable performances. Nonetheless, according to the presented data, it is still fair to deduce that:

- all filters exhibit wide range of tunability

- the achievable speed of the high-speed all-npn log-domain filters can reach the hundred MHz range[†]

- minimum supply voltage of the lowpass log-domain filters can be as low as 1.2V, resulting in less than 1 mW of power consumption.

The trend of circuit evolution toward higher-speed, lower-power implementations is quite obvious here. In fact, as the research of log-domain filters has by no means reached maturity, it is expected that many exciting breakthroughs on integrator circuits will happen in the future.

---

†. Notice that the HSLP log-domain filter is implemented on a quick turn-around, semi-custom bipolar array with no option of linear on-chip capacitor [46]. Therefore, the capacitors have to be kept off-chip, which cannot be too small. As a esult, the reported operating speed is limited to a few MHz's. The full high-speed potential of the circuit has yet to be fully explored.

A Brief Summary of Log-Domain Filter Experimental Results	Basic Design (Section 2.1)	Types of Log-Domain Integrator Design					
		High-Speed (HS) Design (Section 2.3.1)			Low-Power (LP) Design (Section 2.3.2)		HSLP Design (Section 2.3.3)
References	[30]	[17]	[42]	[42]	[44]	[45]	[46]
Technology f_T	2.5 GHz (npn) 10 MHz (pnp)	10.2 GHz (npn) 4.6 GHz (pnp)	11 GHz (npn)	11 GHz (npn)	6 GHz (npn)	6 GHz (npn)	2.5 GHz (npn) 10 MHz (pnp)
Filter Order/Type	5th, lowpass Chebyshev	2nd, bandpass/ lowpass	2nd, bandpass/ lowpass	4th, bandpass maximally flat	3rd, lowpass Chebyshev	3rd, lowpass Chebyshev	3rd, lowpass Chebyshev
Diff. / Single-ended	single-ended	single-ended	single-ended	differential	differential	differential	single-ended
Class A/ AB	Class A	Class A	Class A	Class A	Class A	Class AB	Class A
Tuning Range	up to 1 MHz	125- 430 MHz	83- 220 MHz	50- 130 MHz	10 k - 100 MHz	10 k- 15 MHz	40 k- 4 MHz
Center/ Cutoff Freq	50 kHz	-	-	130 MHz	320 kHz	320 kHz	1.5 MHz
Supply Voltage	±5 V	2.7 V	5V	5 V	1.2 V	1.2 V	1.2 V
Bias Current	180 µ A	0.44- 1.6 mA	0.79- 2.2 mA	220 µ A	1 µ A	1 µ A	25 µ A
Total Power	-	27- 81 mW	90- 235 mW	233 mW	23 µ W	65 µ W	844 µ W
Linearity	THD= -47 dB IMD3= -55 dB	input-referred IP3 = -12.5 & -14 dBm @ f_o= 83 & 400 MHz	IMD3 = -34.73 & -32.5 dB @ f_o= 83 & 220 MHz	IMD3= -45.6 dB @ f_o= 83 MHz	DR= 57 dB @ THD ≤ -40dB	DR= 65 dB @ THD ≤ -40dB	DR= 41 dB @ -40dB HD3

Table 2.1 A brief summary of experimental performances from log-domain filters featuring different integrator circuits

CHAPTER 3

Log-Domain Filter Synthesis-I: Operational Simulation of *LC* Ladders

At this point in the text, we have seen several types of log-domain integrators, each meeting different performance criteria. We have also seen in Chapter 1 a simple linearization procedure for interconnecting nonlinear blocks together to form a linear input-output transfer function. We are now in a position to create arbitrary order filter circuits having desired frequency responses using these log-domain integrator circuits. We shall refer to this procedure as log-domain filter synthesis.

Traditionally speaking, synthesis of analog filters can be achieved in many different ways. Among them, the method of operational simulation of *LC* ladder network prototypes and the method of state-space synthesis have gained widespread use. As we shall see, the synthesis of log-domain filters is no different than the schemes that are used for the synthesis of active-RC, g_m-C or MOS-C filters except for a few minor modifications. This chapter will describe the operational simulation method. The next chapter will outline the synthesis method based on the state-space formulation. Circuit examples and simulation results will be presented throughout these two chapters to illustrate these methods.

3.1 Traditional Active Filter Synthesis by *LC* Ladder Signal Flow Graph

One of the most popular filter synthesis techniques is the operational simulation of lossless doubly terminated *LC* ladder networks. It can also be viewed as a method that stems from a particular signal-flow graph (SFG) representation of the LC ladder. The benefits are many-fold. Widely accepted by the design community, lossless doubly terminated LC ladders that are designed to deliver maximum possible power to the load exhibit a low passband sensitivity to the inevitable process and element variations. The resulting circuit, commonly known as a leapfrog filter, has a one-to-one correspondence with its passive LC ladder counterpart. This enables one to make use of the enormous wealth of knowledge accumulated on LC ladders over the past half-century. We shall make use of this knowledge as a means of analyzing the behavior of log-domain filter circuits subject to nonideal transistor parasitics, as well as suggesting ways to compensate for their effects.

The operational simulation of LC ladders involves finding an active circuit that will mimic the voltage and current relationships of the L and C elements within the LC ladder. The designer first writes a set of integral equations that completely describes the operation of the LC ladder. Then, a SFG representation of these equations is drawn. Once established, the SFG can be implemented in any of a number of technologies such as active-RC, MOS-C and g_m-C using an integrator as the basic building block.

Let us illustrate this method using a 3rd-order lowpass active-RC example. First, an LC ladder prototype that meets the desired attenuation specifications must be found. This can be done through hand analysis [3], by using look-up tables [3], [48], or by using a filter design program [49]-[50]. For illustration purposes, we assume that the third-order LC ladder network shown in Figure 3-1(a) meets the given frequency specifications.

Next, we write a set of equations that completely describes the LC ladder. Here we make use of the circuit analysis method of modified nodal analysis (MNA) [51]. Each capacitor node is assigned a voltage variable and each inductor is assigned a current variable. As these variables are related to the energy storage elements of the system, they are commonly referred to as state-variables. In Figure 3-1(a) we denote the voltages across capacitors c_1 and c_3 as V_1 and V_3, respectively, and the single inductor current as I_2. We also denote the output voltage V_o equal to V_3. Applying KCL to capacitor node V_1, we obtain an integral equation that relates the capacitor node voltage to itself and the other state variables as

$$V_1 = \frac{1}{c_1} \cdot \int \left(\frac{V_S}{r_S} - I_2 - \frac{V_1}{r_S} \right) dt \tag{3.1}$$

Next, we write an impedance description for the inductor in terms of the node voltages as follows

$$I_2 = \frac{1}{l_2} \cdot \int (V_1 - V_3) dt \tag{3.2}$$

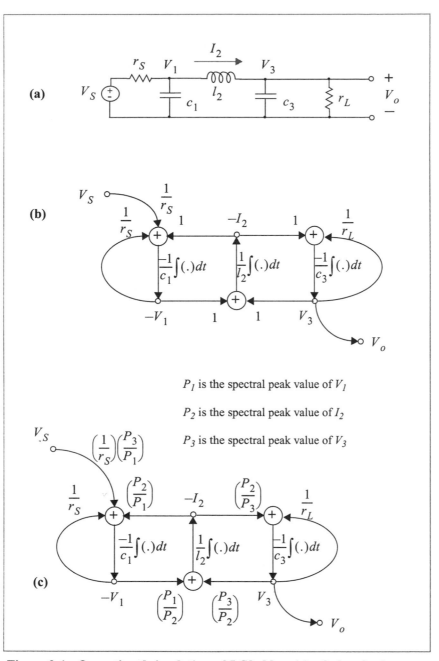

Figure 3-1: Operational simulation of *LC* ladder: (a) a 3rd-order lowpass *LC* ladder filter prototype, (b) the corresponding SFG, (c) scaling the SFG branches for maximum dynamic range, (d) an active RC multiple-input integrator, and (e) the complete active-RC filter. (*continued on next page*)

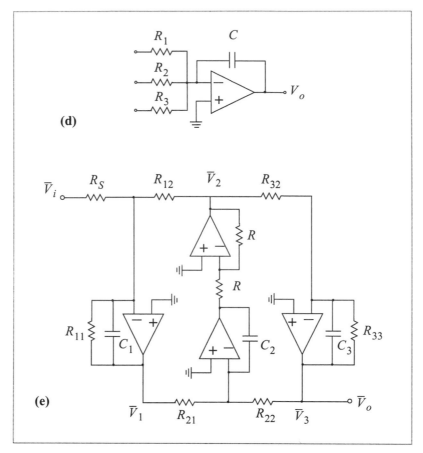

Figure 3-1: (*continued*) **Operational simulation of *LC* ladder**

Finally, we write KCL at capacitor node V_3 to obtain

$$V_3 = \frac{1}{c_3} \cdot \int \left(I_2 - \frac{V_3}{r_l} \right) dt \qquad (3.3)$$

These three equations completely describe the *LC* ladder prototype. We can then draw the SFG based on them and it is shown in Figure 3-1(b).

The next operation that we need to take care of is to scale the initial branch values of the SFG so that the resulting active filter has the widest possible dynamic range [3]. Dynamic range scaling is usually achieved by equalizing the peaks obtained at the outputs of the integrators as the frequency of the input sinusoid is swept over a desired band. To perform scaling one needs the values of the signal peaks, denoted here as P_1, P_2 and P_3 corresponding to the peaks of V_1, I_2 and V_3

in Figure 3-1(c). They can be found from a simulation of the SFG, or alternatively, from an analysis of the LC ladder network.

Once the spectral peaks have been determined, the branch gains of the SFG can be scaled as follows: the branch gain K_{ij} connected between the output of the j-th integrator and the input of the i-th integrator is changed to $K_{ij} \cdot (P_j / P_i)$. As an example, consider that the gain from the 2nd integrator output to the input of the 1st integrator in Figure 3-1(b) changes from the initial value of unity to (P_2 / P_1). Note that the scaling process preserves the magnitude of loop transmission of every two-integrator loop, and hence the filter transfer function (except for a gain constant) remains unchanged. To keep the overall gain unchanged, either the input or output branch should be scaled by the factor (P_3 / P_1). In this particular case, we chose the input branch as no new circuitry is required. Thus the input branch having gain $1/r_s$ is scaled by the factor (P_3 / P_1). In log-domain circuits, we usually find the output branch more suitable for scaling. We summarize this scaling operation in Figure 3-1(c) for the 3rd-order SFG. Extension to other filter orders should be directly apparent.

The final step is to implement the SFG using active-RC blocks such as the multi-input integrator shown in Figure 3-1(d). A corresponding integrator circuit subsequently replaces each branch of the SFG. The result is the circuit shown in Figure 3-1(e). The actual details of this transformation are not important at this point in our discussion. Shortly, we will go to great lengths describing how log-domain integrators are used to realize a particular SFG. We should point out here that the variables in the active-RC circuit are not the same as those in the SFG. However, they do have a one-to-one correspondence. Specifically, the variables in the SFG correspond to those in the active-RC circuit (variables with an overhead bar) as follows:

$$V_S \Leftrightarrow \overline{V}_i \qquad -V_1 \Leftrightarrow \overline{V}_1 \qquad -I_2 \Leftrightarrow \overline{V}_2 \qquad V_3 \Leftrightarrow \overline{V}_3 \qquad V_o \Leftrightarrow \overline{V}_o$$

To summarize, the synthesis method involves the following four steps:

1. Find an LC ladder which meets the desired filter specification

2. Draw the SFG that corresponds to the LC ladder

3. Scale the SFG for maximum dynamic range

4. Implement the SFG using integrators from the desired technology.

It can be observed that up to and including step 3, the synthesis steps are technology-independent. From the same SFG, a number of different circuit realizations, such as active-RC, MOS-C, g_m-C etc., can be obtained. Although the log-domain integrators are non-linear in nature, it is instructive to see how they can be derived from the same

SFG.

3.2 Log-Domain Ladder Filter Synthesis

Beginning with a linear signal-flow graph, log-domain filters are constructed by first modifying the SFG so that it contains nonlinear branches that can be realized by a set of log-domain integrators. In short, log-domain filter synthesis using the method of operational simulation of *LC* ladder networks involves the following four steps [29]-[30]:

1. Find an *LC* ladder that meets the design specifications

2. Derive the corresponding SFG from the *LC* prototype

3. Modify the SFG to obtain the log-domain equivalence by:

    a. placing a *LOG* block after each integrator

    b. placing an *EXP* block at the input to each summer (before
       the multiplicative factor)

    c. placing an *EXP* block at the output of the system

    d. placing a *LOG* block at the input to the system

4. Replace the integrator branches of the log-domain SFG with log-domain
   integrator circuits.

Notice that by performing step 3, we are transforming the filter SFG from the linear-domain into the log-domain, while ensuring overall input-to-output linearity by the linearization techniques discussed in Section 1.4.4. This procedure is illustrated in Figure 3-2. By examining the final log-domain SFG we can see how the overall linear transfer function has been maintained. A *LOG* block immediately follows each *EXP* block. Therefore, due to the complementary nature of these functions, the linearity and the integrity of the original system has been conserved. The areas enclosed by dashed boxes represent typical log-domain building blocks, which conform to the paradigm of Figure 1-10 (or more specifically, the integrator of Figure 2-3). Note also the presence of the *LOG* block at the input and the *EXP* block at the output to the system. We refer to the variables in the log-domain with a circumflex (^). The input and output variables are linear variables so they do not have a circumflex placed over them. We shall maintain this notation throughout the rest of this text.

Using the basic log-domain integrators of Section 2.1, four log-domain filter examples will be presented based on the operational simulation method. They are respectively the lowpass/bandpass biquadratic/high-order log-domain filters. For illustrative purposes, they will be synthesized with the basic single-ended log-domain

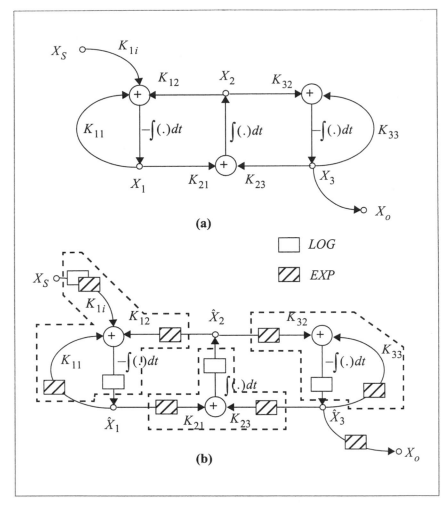

Figure 3-2: (a) linear SFG of a typical *LC* ladder, and (b) the equivalent log-domain SFG.

integrator of Figure 2-2. It is selected because of its simplicity and sufficiency to demonstrate all the fundamental principles. It may not necessarily yield the highest performances with respect to speed, linearity, and noise and power consumption as discussed before. Nevertheless, one should note that the log-domain filter synthesis examples demonstrated here are equally applicable to all integrator circuits presented in Chapter 2. This can be done even though these integrators implement different *LOG* and *EXP* functions. In other words, we can routinely replace the basic integrator circuits of the log-domain filter examples (to be presented shortly) with any of the more sophisticated designs such as those depicted in Figures 2-7 to 2-11, given a minor modification of the capacitor scaling factor[†].

3.2.1 Log-Domain Lowpass Biquadratic Filter

In this subsection, we shall illustrate the design of a log-domain lowpass biquadratic filter. The steps associated with the synthesis of a log-domain lowpass biquadratic filter is depicted graphically in Figure 3-3. It begins by first finding a passive LC ladder prototype that meets the desired frequency specifications, such as the second-order LC ladder network shown in Figure 3-3(a). This LC ladder implements the following transfer function

$$\frac{V_o}{V_i}(s) = \frac{1/(lc)}{s^2 + s\left(\frac{1}{rc}\right) + \frac{1}{lc}} \tag{3.4}$$

When compared with the transfer function written in standard form, i.e.,

$$H(s) = K \cdot \frac{\omega_o^2}{s^2 + \left(\frac{\omega_o}{Q}\right)s + \omega_o^2} \tag{3.5}$$

the cutoff frequency, filter Q factor, and the dc gain are

$$\omega_o = \frac{1}{\sqrt{lc}} \qquad Q = r \cdot \sqrt{\frac{c}{l}} \qquad K = 1 \tag{3.6}$$

Next, we write a set of equations that completely describes the LC ladder. This requires two equations; one KVL equation relating the inductor current I_1 to its terminal voltages as,

$$I_1 = \frac{1}{l} \cdot \int (V_i - V_2) dt \tag{3.7}$$

---

†. For instance, comparing the log-domain integrator equations of the basic (Figure 2-2) and the all-npn (Figure 2-7) structures (which are recaptured from (2.5) and (2.12) respectively):

basic integrator: $\quad EXP(\hat{V}_o) = \frac{I_o}{2V_T} \cdot \frac{1}{C} \cdot \int \{EXP(\hat{V}_{ip}) - EXP(\hat{V}_{in})\} dt$

all npn integrator : $\quad EXP(\hat{V}_o) = \frac{I_o}{V_T} \cdot \frac{1}{C} \cdot \int \{EXP(\hat{V}_{ip}) - EXP(\hat{V}_{in})\} dt$

it can be seen that their integration scalar factors differ by a factor of two. Therefore, in order to achieve an identical filter transfer function under the same bias current, the capacitance required for the all npn realization will be twice the values of those required by the basic structure.

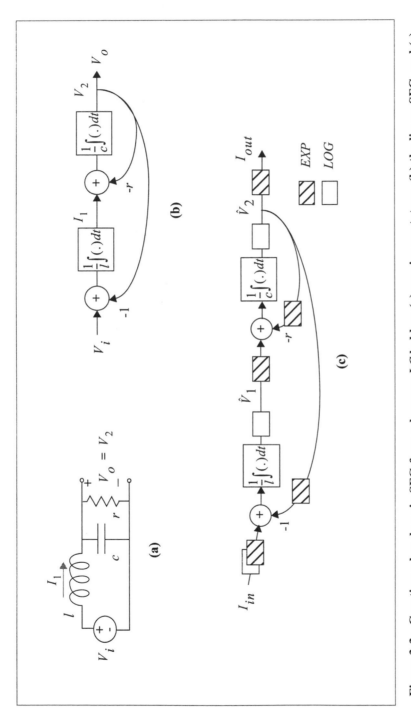

Figure 3-3: Creating a log-domain SFG from a lowpass *LC* ladder: (a) passive prototype, (b) the linear SFG, and (c) the corresponding log-domain SFG.

and the other from KCL equation that relates the capacitor node voltage V_2 to the other state-variables as

$$V_2 = \frac{1}{c} \cdot \int \left(I_1 - \frac{V_2}{r} \right) dt \qquad (3.8)$$

These two equations completely characterize the *LC* ladder prototype. We can now draw the SFG based on them as shown in Figure 3-3(b). Routinely adding the *LOG* and *EXP* cells to the linear SFG, the log-domain SFG is obtained as shown in Figure 3-3(c). The variables in the two SFGs correspond according to the following mappings:

$$V_i \Leftrightarrow I_{in} \qquad I_1 \Leftrightarrow \hat{V}_1 \qquad V_2 \Leftrightarrow \hat{V}_2 \qquad V_o \Leftrightarrow I_{out}$$

Finally, the log-domain filter circuit is obtained by replacing the appropriate branches of the log-domain SFG with multiple-input log-domain integrators and inserting at the input a *LOG* circuit and at the output an *EXP* circuit. We depict this process for each integrator section of the log-domain SFG in Figures 3-4 and 3-5, together with a further circuit simplification. For example, the SFG representing the first integrator is shown in Figure 3-4(a). It consists of two feed-in branches; one connected to the input I_{in} and the other to the node $\hat{V}_2$. Correspondingly, a log-domain circuit that has a similar form as this SFG is shown in Figure 3-4(b). It consists of transistors Q_1-Q_4 and Q_1'-Q_4' making up the input branch with bias current I_{1i}. Likewise, transistors Q_9-Q_{12} and Q_9'-Q_{12}' with bias current I_{12} make up the integration branch from $\hat{V}_2$. Writing KCL at the capacitor node, we obtain

$$C_1 \cdot \frac{d\hat{V}_1}{dt} = \left[(I_{1i} + I_{in}) e^{\frac{-\hat{V}_1}{2V_T}} - I_{1i} e^{\frac{-\hat{V}_1}{2V_T}} \right] + \left[I_{12} e^{\frac{-\hat{V}_1}{2V_T}} - I_{12} e^{\frac{\hat{V}_2 - \hat{V}_1}{2V_T}} \right] \qquad (3.9)$$

or on simplifying

$$C_1 \cdot \frac{d\hat{V}_1}{dt} = I_{in} e^{\frac{-\hat{V}_1}{2V_T}} + I_{12} e^{\frac{-\hat{V}_1}{2V_T}} - I_{12} e^{\frac{\hat{V}_2 - \hat{V}_1}{2V_T}} \qquad (3.10)$$

Following a similar procedure to that outlined in Section 2.1.1, we multiply both sides by $\exp(\hat{V}_1 / 2V_T)$ and re-arrange to obtain

$$\frac{2V_TC_1}{I_o} \cdot \frac{d}{dt}\left(I_o e^{\frac{\hat{V}_1}{2V_T}} - I_o\right) = I_{in} - \frac{I_{12}}{I_o}\left(I_o e^{\frac{\hat{V}_2}{2V_T}} - I_o\right) \quad (3.11)$$

Making use of the familiar *LOG* and *EXP* mapping functions,

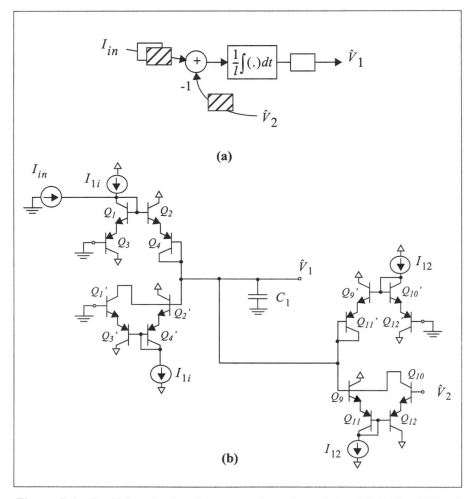

(a)

(b)

Figure 3-4: Realizing the first integrator branches of the *LC* ladder SFG of Figure 3-3 with log-domain cells: (a) first integrator SFG branches, and (b) replacing each branch of the SFG with log-domain cells.

$$LOG(x) = 2V_T \ln\left(\frac{I_o + x}{I_o}\right) \qquad EXP(x) = I_o e^{\frac{x}{2V_T}} - I_o, \qquad (3.12)$$

we obtain

$$EXP(\hat{V}_1) = \frac{I_o}{2V_T} \cdot \frac{1}{C_1} \cdot \int \left[I_{in} - \frac{I_{12}}{I_o} EXP(\hat{V}_2)\right] dt \qquad (3.13)$$

Comparing this expression with the one derived from the SFG in Figure 3-4(a), i.e.,

$$EXP(\hat{V}_1) = \frac{1}{l} \cdot \int [I_{in} - EXP(\hat{V}_2)] dt \qquad (3.14)$$

we obtain an equivalence between the SFG and the log-domain circuit, if and only if,

$$C_1 = \frac{I_o}{2V_T} \cdot l \qquad (3.15)$$

and

$$I_{12} = I_o. \qquad (3.16)$$

In the above example, the SFG was not scaled for maximum dynamic range. If we were to include the appropriate dynamic range scale factors then (3.15) and (3.16) would include these scale factors.

Following a similar procedure to above, Figure 3-5 illustrates the construction of the second integration stage whose SFG is depicted in Figure 3-5(a). The corresponding log-domain circuit is shown in Figure 3-5(b) whose output voltage $\hat{V}_2$ is described by the following equation

$$C_2 \cdot \frac{d\hat{V}_2}{dt} = \left[I_{21} e^{\frac{\hat{V}_1 - \hat{V}_2}{2V_T}} - I_{21} e^{\frac{-\hat{V}_2}{2V_T}}\right] + \left[I_{22} e^{\frac{-\hat{V}_2}{2V_T}} - I_{22} e^{\frac{\hat{V}_2 - \hat{V}_2}{2V_T}}\right] \qquad (3.17)$$

Multiplying both sides by $e^{\hat{V}_2/2V_T}$ and replacing $e^{(\hat{V}_2 - \hat{V}_2)/2V_T}$ by unity, we can write

$$\frac{2V_T C_2}{I_o} \cdot \frac{d}{dt}\left(I_o e^{\frac{\hat{V}_2}{2V_T}} - I_o\right) = \frac{I_{21}}{I_o}\left(I_o e^{\frac{\hat{V}_1}{2V_T}} - I_o\right) - \frac{I_{22}}{I_o}\left(I_o e^{\frac{\hat{V}_2}{2V_T}} - I_o\right) \qquad (3.18)$$

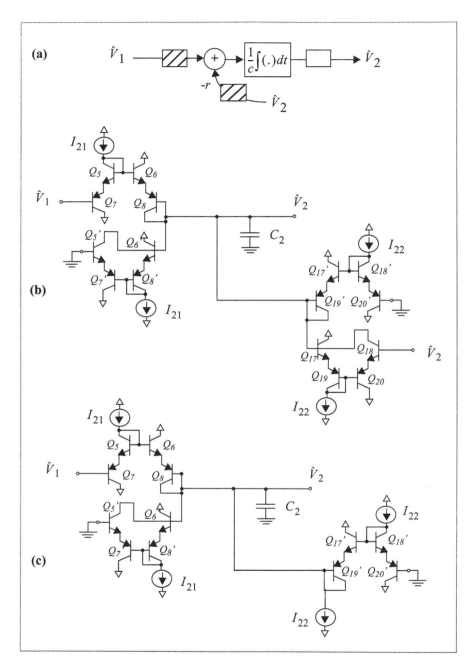

Figure 3-5: Realizing second integrator branches of the LC ladder SFG of Figure 3-3: (a) second integrator SFG branches, (b) replacing each branch of the SFG with log-domain cells, and (c) replacing damping cell with equivalent current source.

Re-arranging (3.18) using the mappings in Eq. (3.12), we get

$$EXP(\hat{V}_2) = \frac{I_o}{2V_T} \cdot \frac{1}{C_2} \cdot \int \left[\frac{I_{21}}{I_o} EXP(\hat{V}_1) - \frac{I_{22}}{I_o} EXP(\hat{V}_2) \right] dt \qquad (3.19)$$

Comparing this expression with the one derived from the SFG in Figure 3-5(a), i.e.,

$$EXP(\hat{V}_2) = \frac{1}{c} \cdot \int \left[EXP(\hat{V}_1) - \frac{1}{r} EXP(\hat{V}_2) \right] dt \qquad (3.20)$$

we obtain

$$C_2 = \frac{I_o}{2V_T} \cdot c \qquad I_{21} = I_o \qquad I_{22} = \frac{I_o}{r} \qquad (3.21)$$

One can further simplify the circuit in Figure 3-5(b) by replacing the log-cell consisting of transistors Q_{17}-Q_{20} whose input connects directly back to its output by a constant current source as shown in Figure 3-5(c). Such a substitution was described in Section 2.1.2.

Finally, interconnecting the two integrating stages and adding an output EXP stage (i.e., transistors Q_{13}-Q_{16} biased at a current level of $K \cdot I_o$), we arrive at the circuit shown in Figure 3-6. This active realization mimics the behavior of the passive second-order LC circuit with a DC gain of K. The current ratio I_{out}/I_{in} will ideally realize the transfer function of (3.5) with parameters described by (3.6). Its behavior subject to physical frequencies, i.e. $s = j\omega$, can then be written as

$$\frac{I_{out}}{I_{in}} = H(\omega) = K \cdot \frac{1}{1 - \left(\frac{\omega}{\omega_o} \right)^2 + j \left(\frac{\omega}{\omega_o} \right) \frac{1}{Q}} \qquad (3.22)$$

It is interesting to note that the circuit of Figure 3-6 can be greatly reduced if we impose a simple condition on the LC ladder network in Figure 3-3(a). Specifically, if we eliminate one of the three degrees of freedom used to define the second-order transfer function by setting $r = 1$, then each integrator in the SFG of Figure 3-3(b) will contain only one parameter controlling its gain. The result is that all the log-cells will be biased at the same current level of I_o. This in turn leads to the cancellation of all pairs of positive and negative log-cells whose inputs are grounded. For instance, in Figure 3-6 with $I_{11} = I_{12} = I_o$, the positive current supplied by the log-cell composed of transistors Q_9'-Q_{12}' would cancel with the current pulled by the log-cell composed of transistors Q_1'-Q_4'. Hence, these log-cells can be eliminated

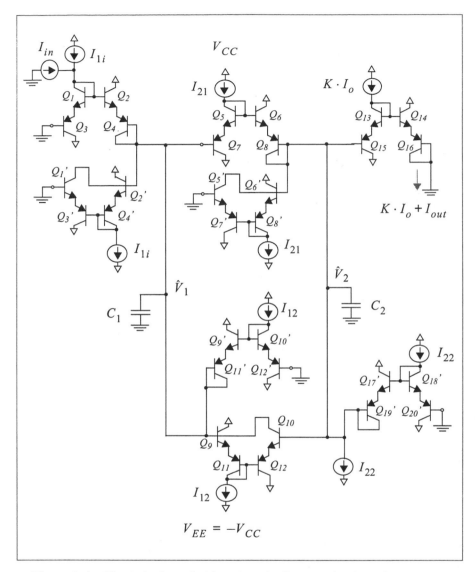

Figure 3-6: **Final single-ended log-domain filter realization of the second-order lowpass *LC* ladder of Figure 3-3.**

from the circuit [70], leading to possible circuit improvement from potential noise sources and parasitic capacitances. Likewise, the grounded input cells made up of transistors of Q_5'-Q_8' and Q_{17}'-Q_{20}' can be eliminated. The result is the simplified circuit shown in Figure 3-7. The drawback to this circuit is that it cannot be scaled for maximum dynamic range. Interesting enough, this circuit is identical to the one first proposed by Frey using the exponential state-space synthesis method [22]. We have more to say about this method in the first section of the next chapter.

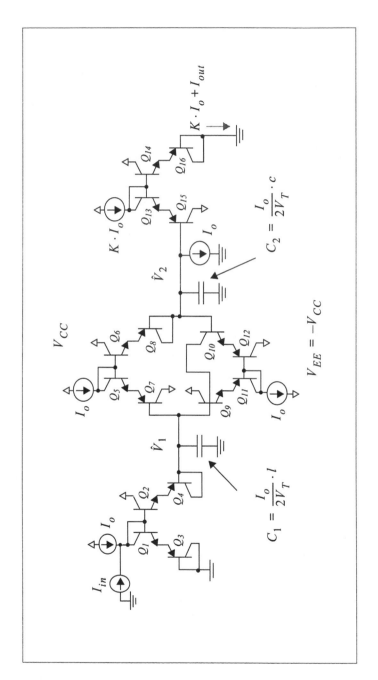

Figure 3-7: Simplified log-domain filter realization of Figure 3-6.

3.2.2 Log-Domain Bandpass Biquadratic Filter

By means of a synthesis procedure similar to the one used in the previous example [31], the bandpass biquad filter is realized from an LC prototype as shown in Figure 3-8(a). It implements a transfer function of the form

$$\frac{I_{out}}{I_{in}} = H(s) = K \cdot \frac{\left(\frac{\omega_o}{Q}\right)s}{s^2 + \left(\frac{\omega_o}{Q}\right)s + \omega_o^2} \tag{3.23}$$

where ω_o, Q and K are identical to that given in (3.6) with $r=1$, where K is the center frequency gain. Expressed in terms of physical frequencies $s = j\omega$, (3.23) becomes

$$H(\omega) = K \cdot \frac{j\left(\frac{\omega}{\omega_o}\right)\frac{1}{Q}}{1 - \left(\frac{\omega}{\omega_o}\right)^2 + j\left(\frac{\omega}{\omega_o}\right)\frac{1}{Q}} \tag{3.24}$$

which is a conventional bandpass filter response. The corresponding SFG representation is shown in Figure 3-8(b) and its equivalent log-domain SFG is provided in Figure 3-8(c). Following the procedure of the last example, Figures 3-9 and 3-10 depict the development of the first and second integrator stage, respectively. Of particular interest here is the simplification of the log-domain circuit shown in Figures 3-9(c). In this case, not all the grounded input log-domain cells are eliminated, one still remains. Here we have chosen the log-cell created by transistors Q_1'-Q_4', although the grounded-input log-cells created by Q_9'-Q_{12}' could have been used instead. The final circuit is shown in Figure 3-11 with all redundant log-domain cells eliminated. Here the bandpass output is derived from the node voltage $\hat{V}_1$ using an *EXP* output stage consisting of transistors Q_{13}-Q_{16}. A lowpass output is also available if the output is derived from $\hat{V}_2$ instead [31].

Before leaving this subsection, we should point out that by canceling the log-domain cells, the circuit of Figure 3-11 cannot be scaled for maximum dynamic range. The remaining examples of this chapter will also forfeit this ability to scale in order to create simpler circuits. Nonetheless, by returning to the original integrator (consisting of both positive and negative log-cells), this feature can be re-established.

3.2.3 High-Order Log-Domain Lowpass Filter

In this case, we wish to design a 7th-order log-domain Chebyshev lowpass filter that meets the following specifications:

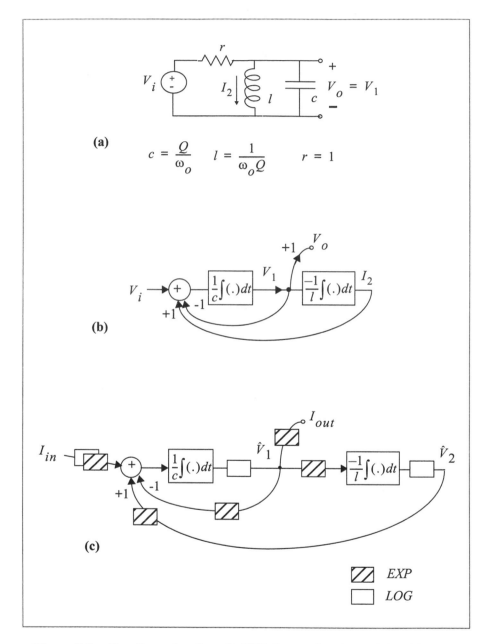

Figure 3-8: Creating a log-domain SFG from a bandpass LC ladder: (a) passive prototype, (b) the linear SFG, and (c) the corresponding log-domain SFG.

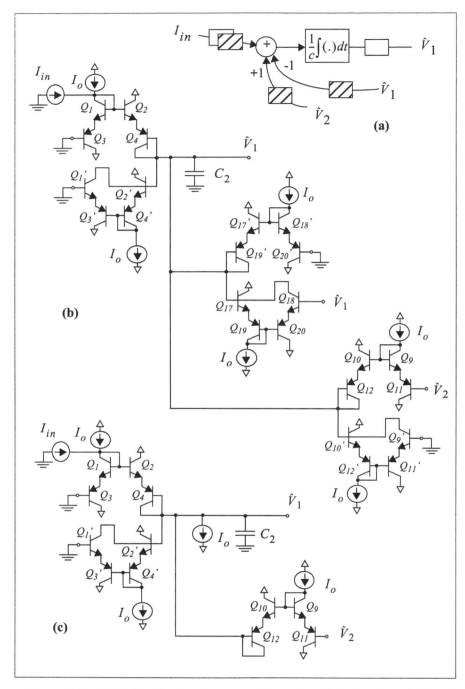

Figure 3-9: Realizing the first integrator branches of the log-domain band-pass biquad: (a) the log-domain SFG, (b) log-domain circuit implementation, and (c) the simplified form.

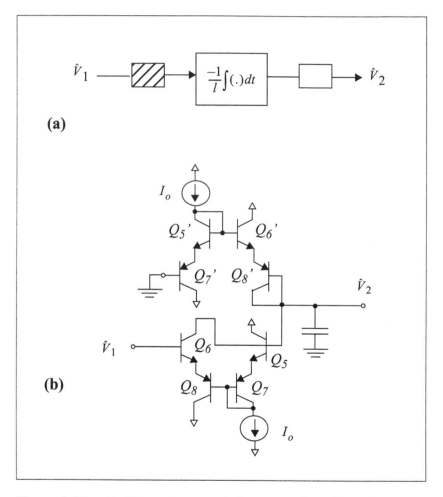

Figure 3-10: Realizing the second integrator branches of the log-domain bandpass biquad: (a) the log-domain SFG, (b) the log-domain integrator implementation.

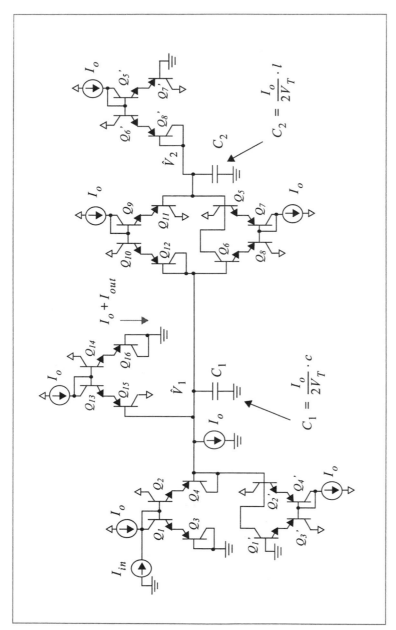

Figure 3-11: Final log-domain filter realization of the second-order bandpass *LC* ladder of Figure 3-8.

Cutoff Frequency = 1 MHz

Passband Ripple = 1 dB

Figure 3-12 shows the LC ladder prototype. With the help of a filter design program [49], the component values are found and summarized in Table 3-1.

To obtain the SFG corresponding to the LC ladder we apply the method of modified nodal analysis. Each capacitor node is assigned a voltage variable (V_1, V_3, V_5, V_7), and each inductor is given a current variable (I_2, I_4, I_6). This signal assignment is shown in Figure 3-12. We then write the impedance description for each inductor in terms of its node voltages and the inductor current. This results in the following three equations:

$$I_2 = \frac{1}{l_2} \cdot \int (V_1 - V_3) dt \tag{3.25a}$$

$$I_4 = \frac{1}{l_4} \cdot \int (V_3 - V_5) dt \tag{3.25b}$$

$$I_6 = \frac{1}{l_6} \cdot \int (V_5 - V_7) dt \tag{3.25c}$$

Next, we apply KCL to the previously labeled voltage nodes and derive a set of equations relating the node voltages, inductor currents and the input. For the LC ladder of Figure 3-12, the following equations are written:

$$V_1 = \frac{1}{c_1} \cdot \int \left(\frac{V_i - V_1}{1} - I_2 \right) dt \tag{3.25d}$$

$$V_3 = \frac{1}{c_3} \cdot \int (I_2 - I_4) dt \tag{3.25e}$$

Name	Value	Name	Value
c_1	0.3448 μF	l_2	0.1769 μH
c_3	0.4924 μF	l_4	0.1868 μH
c_5	0.4924 μF	l_6	0.1769 μH
c_7	0.3448 μF		

Table 3-1: Component values for the 7th-order lowpass LC ladder

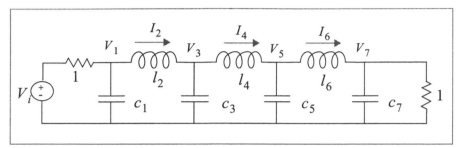

Figure 3-12: 7th-order Chebyshev lowpass *LC* ladder prototype.

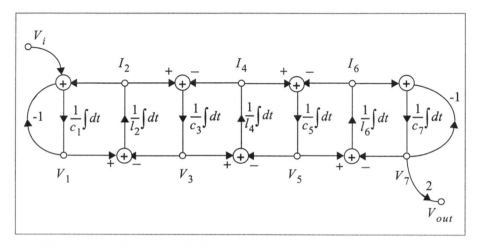

Figure 3-13: 7th-order Chebyshev lowpass *LC* ladder prototype.

$$V_5 = \frac{1}{c_5} \cdot \int (I_4 - I_6) dt \qquad (3.25f)$$

$$V_7 = \frac{1}{c_7} \cdot \int \left(I_6 - \frac{V_7}{1} \right) dt \qquad (3.25g)$$

The doubly terminated *LC* ladder network shown in Figure 3-12 has a DC gain of 0.5. In order to compensate for this loss, the voltage across the load resistor is usually scaled by a factor of 2. Hence, we write

$$V_{out} = 2 \cdot V_7 \qquad (3.26)$$

This set of equations completely specifies the *LC* ladder prototype. Based on it, we can now draw the corresponding SFG as shown in Figure 3-13. Routinely adding the *LOG* and *EXP* blocks according to the rules presented before, and after the following mappings are made,

$$V_i \Leftrightarrow I_{in} \quad V_1 \Leftrightarrow \hat{V}_1 \quad V_3 \Leftrightarrow \hat{V}_3 \quad V_5 \Leftrightarrow \hat{V}_5 \quad V_7 \Leftrightarrow \hat{V}_7$$

$$I_2 \Leftrightarrow \hat{V}_2 \quad I_4 \Leftrightarrow \hat{V}_4 \quad I_6 \Leftrightarrow \hat{V}_6 \quad V_{out} \Leftrightarrow I_{out}$$

(3.27)

the log-domain SFG shown in Figure 3-14(a) results. Note how the operation of the original SFG has been maintained due to the inverse nature of the *LOG* and *EXP* functions.

The final step in the design of log-domain filter involves replacing the different components of the SFG of Figure 3-14(a) with the log-domain integrator circuit. This results in the complete log-domain circuit shown in Figure 3-14(b), where $\{C_1, C_2, ..., C_7\}$ correspond to $I_o/(2V_T) \cdot \{c_1, l_2, c_3, l_4, c_5, l_6, c_7\}$, respectively. In addition, the output *EXP* stage is biased at $2I_o$ in order to account for the DC gain of the *LC* ladder.

3.2.4 High-Order Log-Domain Bandpass Filter

The last example to be demonstrated is a 6th-order Chebyshev bandpass filter. To begin with, the *LC* ladder prototype is shown in Figure 3-15. The component values, which give a 1 MHz center frequency and 1 dB passband ripple, are tabulated in Table 3-2. The log-domain filter synthesis procedures are very similar to the previous lowpass case. Therefore, we will go over them quickly.

Again, by modified nodal analysis, a voltage variable (V_1, V_2, V_3) is assigned across each capacitor and a current variable (I_4, I_5, I_6) is assigned to each inductor. This is demonstrated in Figure 3-15.

Straightforward analysis will arrive at the following inter-relationships for the voltage and current variables:

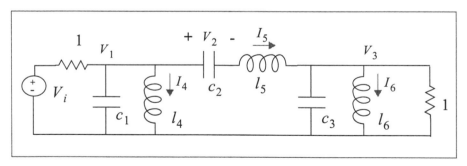

Figure 3-15: 6th-order Chebyshev bandpass *LC* ladder prototype.

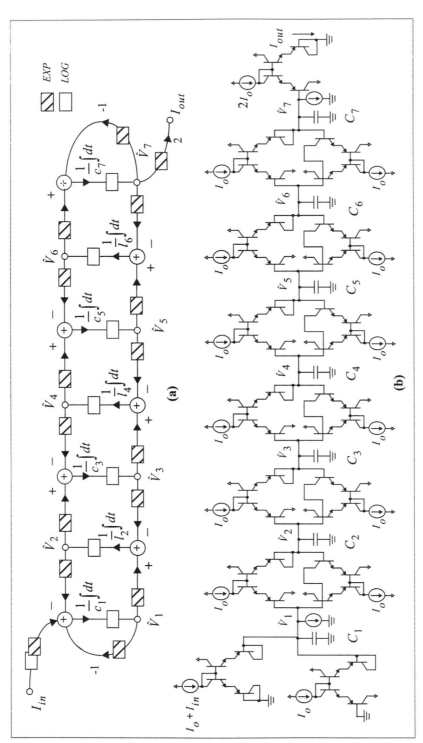

Figure 3-14: 7th-order Chebyshev lowpass filter: (a) the log-domain SFG, (b) the actual circuit implementation.

Name	Value	Name	Value
c_1	3.221 μF	l_4	7.865 nH
c_2	16.001 nF	l_5	1.5822 μH
c_3	3.221 μF	l_6	7.865 nH

Table 3-2: Component values for the 6th-order bandpass *LC* ladder.

$$V_1 = \frac{1}{c_1}\int\left(\frac{V_i - V_1}{1} - I_4 - I_5\right)dt \tag{3.28a}$$

$$I_4 = \frac{1}{l_4}\int(V_1)dt \tag{3.28b}$$

$$V_2 = \frac{1}{c_2}\int(I_5)dt \tag{3.28c}$$

$$I_5 = \frac{1}{l_5}\int(V_1 - V_2 - V_3)dt \tag{3.28d}$$

$$V_3 = \frac{1}{c_3}\int\left(I_5 - I_6 - \frac{V_3}{1}\right)dt \tag{3.28e}$$

$$I_6 = \frac{1}{l_6}\int(V_3)dt \tag{3.28f}$$

This set of equations completely characterizes the signal operations performed by the passive *LC* ladder prototype. We should adjust for the passband gain of the filter. In this particular case, we should scale the output voltage of the filter by a factor of 2. Hence, we write

$$V_{out} = 2 \cdot V_3 \tag{3.29}$$

Collectively, the above set of equations can be equivalently represented by the linear SFG shown in Figure 3-16(a). After the following mappings are made,

$$V_i \Leftrightarrow I_{in} \qquad V_1 \Leftrightarrow \hat{V}_1 \qquad -I_4 \Leftrightarrow \hat{V}_2 \qquad I_5 \Leftrightarrow \hat{V}_3$$
$$-V_2 \Leftrightarrow \hat{V}_4 \qquad V_3 \Leftrightarrow \hat{V}_5 \qquad -I_6 \Leftrightarrow \hat{V}_6 \tag{3.30}$$

the log-domain SFG is achieved as in Figure 3-16(b) when the *LOG* and *EXP* blocks are routinely added. Replacing each part of the log-domain SFG with its corresponding circuit, and eliminating redundant log-cells, we arrive at the final circuit shown in Figure 3-17 where the capacitor values are

$$\{C_1, C_2, ..., C_6\} = \frac{I_o}{2V_T} \cdot \{c_1, l_4, l_5, c_2, c_3, l_6\} \tag{3.31}$$

This completes our sixth-order bandpass log-domain filter example.

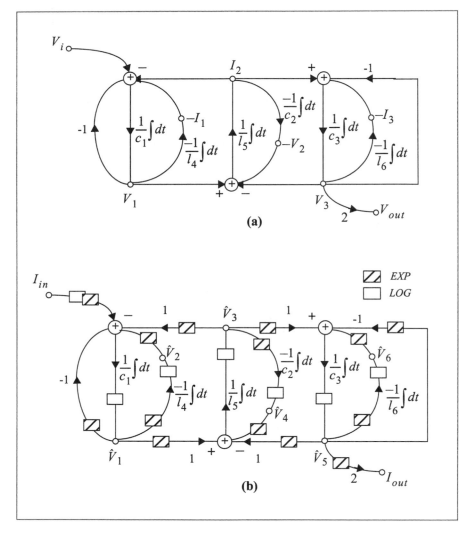

Figure 3-16: Synthesis of sixth-order bandpass log-domain filter: (a) the linear SFG, and (b) the log-domain SFG.

3.2.5 Comments on the Designs

Now that several complete log-domain filters have been designed, we shall take a minute to examine the physical nature of these circuits. There exists a direct correspondence between the variables in the log-domain circuits and the physical voltages and currents in the passive ladder prototype. Such relationships are, for instance, depicted in (3.27) and (3.30). Therefore, it is conceivable that the resulting log-domain ladder filter is closely performing the signal processing of its prototyping *LC* ladder. We can expect the log-domain filters to effectively inherit the advantages offered by classical *LC* ladders, such as their superior passband sensitivity and noise properties. Simulations to be presented later will examine this claim.

It is interesting to note that while voltages represent the input and output signals from the *LC* ladder, the input and output variables from the log-domain filter are both currents. This has led some to characterize the log-domain filter as a *current-mode circuit*. Nonetheless, what has been shown is that key nodes in the log-domain SFG are physically represented by voltages, such as $\hat{V}_1$, $\hat{V}_2$, ..., $\hat{V}_6$ of Figure 3-16(b). In fact, the input and output are only currents due to the *LOG* and *EXP* functions, which have been added to preserve the linearity of the system. This would indicate that this is a voltage-mode circuit. Suffice to say that it is always difficult to classify circuits in this manner due to the fundamental relationship between voltages and currents in a circuit, such as that described by Ohm's law. The important thing is to realize that the final log-domain filters possess many of the advantages that are often associated with current-mode circuits. That includes low impedances along the signal path (suitable for high-speed operation), and very small voltage swings at all key nodes (ideal for low-voltage low-power application).

Before leaving this section, it should be noted that the filter synthesis method presented here is *exact*. If we assume ideal elements with no ohmic resistances, zero base current, and perfect diode relationship, the resulting filter response will be ideal. However in reality, this is hardly the case. Figure 3-18 compares the ideal and the realistic magnitude responses of the four log-domain filter circuits presented before. The simulations were performed using both ideal and nonideal[†] transistor models. The simulation results reveal that actual log-domain filters suffer from center and cutoff frequency shifts, as well as Q-degradation and attenuation of passband ripples. Detailed discussions will be provided in Chapters 5 and 6, where we will analyze and quantify the underlying mechanisms behind these deviations. Besides, compensation schemes will be presented that bring the deviated

---

†. Ideally, bipolar transistor will feature zero ohmic junction resistances, infinite β and Early voltage. For a typical realistic complementary bipolar process, we have assumed the following transistor parameters:

npn: RE= 12 Ω, RB= 1200 Ω, β = 147, V_A= 80 V; and

pnp: RE= 20 Ω, RB= 150 Ω, β = 49, V_A= 60V

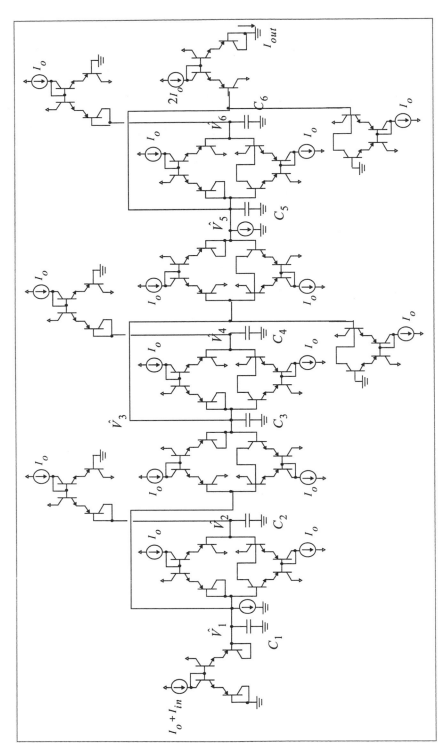

Figure 3-17: 6th-order Chebyshev bandpass log-domain filter.

functions back to their desirable responses.

3.3   Simulation: Large-Signal vs. Small-Signal Analysis

An important question of simulation arises now. Is a small-signal analysis sufficient to fully capture the input-output frequency behavior of log-domain circuits? In general, SPICE is limited by its inability to perform AC analysis on non-linear circuits. This is due to the fact that when performing an AC analysis, SPICE first solves for the DC operating point of the circuit, then determines linearized, small-signal models for all of the non-linear devices in the circuit. Subsequently, the small-signal frequency response over a specified frequency range is computed. As this approach essentially negates the translinear nature of the bipolar transistor arrangement (i.e., eliminates large-signal behavior), it is not ideally suited to testing log-domain circuits. A better approach is to perform a multitone analysis.

Multitone analysis is popular in the testing community and involves finding the spectral response of a circuit stimulated by an input consisting of many sinusoidal tones (a *multitone* input). Appendix B describes the multitone simulation method used in this textbook. This novel approach combines multitone testing with SPICE transient analysis [71]. Because of the large-signal nature of SPICE transient analysis, the circuit is simulated without small-signal approximation. In addition to allowing the measurement of frequency response, multitone analysis can also be used for distortion measurements.

To demonstrate this approach, the 7th-order lowpass Chebyshev filter of Figure 3-14(b) is simulated. The bias current level is set to 100 µA, which corresponds to a 1 MHz cutoff frequency. A signal consisting of twenty equal-amplitude sinewaves is applied as input, which collectively sums up to a peak current of 80 µA (i.e., 80% of bias level, or modulation index=0.8). This safely establishes our large signal input conditions. Evenly spaced in the frequency domain, the twenty sinusoidal tones are chosen to cover a spectrum from dc up to 2.44 MHz (i.e., approximately two times the filter cutoff frequency). SPICE time-domain simulation is then carried out. The resulting output transient waveform is submitted to an FFT (Fast Fourier Transform) analysis so that its frequency response can be extracted.

Figure 3-19(a) and (b) demonstrate the multitone simulation results assuming ideal and realistic transistor models, respectively. The amplitudes of the individual tones agree very closely with the filter response as calculated by the SPICE AC analysis (dotted line).

Multitone analysis is the best method for testing log-domain filters since AC analysis does not account for the translinear nature of these circuits. However, once the operation of the circuit has been confirmed through multitone analysis, AC analysis can then be used as a quicker way to get the frequency response. This is not because the AC analysis results are wrong; it is just that they are limited to small signal inputs.

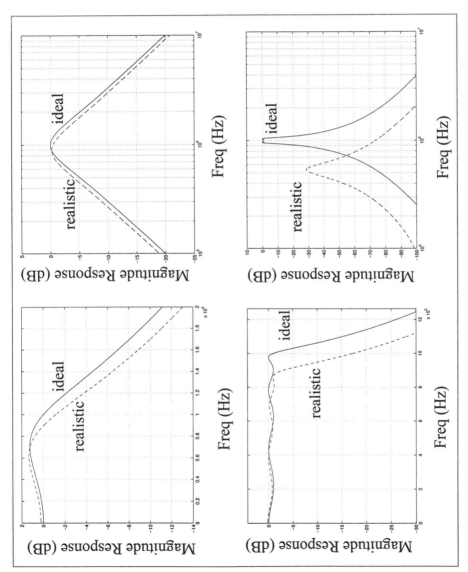

Figure 3-18: Simulations of log-domain filters with ideal and realistic transistor models: (a) lowpass biquad filter, (b) bandpass biquad filter, (c) 7th-order Chebyshev lowpass ladder filter, and (d) 6th-order Chebyshev bandpass filter.

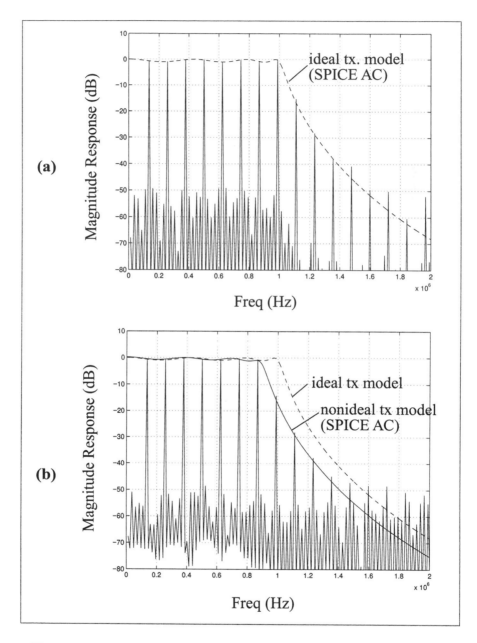

Figure 3-19: SPICE multitone simulation of the log-domain 7th-order low-pass filter with (a) ideal transistor models, and (b) nonideal transistor models. They are compared with the SPICE AC simulated results.

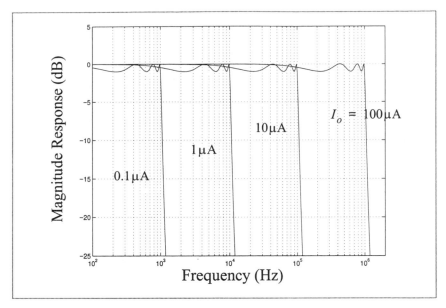

Figure 3-20: Cutoff Frequency tuning of the log-domain filter.

3.4   Electronic Tunability of Log-Domain Filters

As the log-domain technique was first originated in 1979 [21], one of the goals was to achieve easy electronic frequency tunability. In this section, we shall investigate this aspect of log-domain filters.

We have already shown that in order to maintain the equivalence between a log-domain filter and its corresponding linear prototype (LC ladder), the capacitance should be scaled by the factor of $I_o/(2V_T)$. One example is given in the expression of (3.31). Failing to do so will cause an unwanted frequency shift to the resulting log-domain filter behavior. While this adds an extra element of complexity to the synthesis procedure, it is also what makes these circuits easily tunable.

Let us examine the frequency tuning of the 7th-order log-domain lowpass filter in Figure 3-14. The integrating capacitances will be kept constant, while all of the bias currents are varied simultaneously. Figure 3-20 shows the AC analysis results simulated with bias levels of 0.1 µA, 1 µA, 10 µA and 100 µA. The input signal level is chosen such that it is of the same order as the bias current. It can be seen that the filter cutoff frequency is effectively tuned from 1 kHz up to 1 MHz. Figure 3-20 confirms that the filter cutoff frequency is proportional to the bias current, such that

$$\frac{f_c'}{f_c} = \frac{I_o'}{I_o} \qquad\qquad (3.32)$$

while the filter shape (such as the passband ripples) are preserved. This is known as frequency scaling and has an LC ladder counterpart. Specifically, tuning the bias current from I_o to $k \cdot I_o$ in log-domain can be equivalently understood as tweaking all of the reactive components of an LC ladder from $\{l, c\}$ to $1/k \cdot \{l, c\}$. More discussions will be provided on this property in later chapters.

Due to the frequency response dependence on bias current, decades of cutoff frequencies can be programmed and provided by the same log-domain circuit. However, one clear disadvantage is that *any* fluctuations in the bias currents will lead to changes in the cutoff frequency. The solution is to use a tuning scheme that adaptively adjusts the bias level such that the desired cut-off frequency locks to a pre-defined value.

A closely related problem associated with the bias current and its effect on the log-domain circuit is its temperature sensitivity. Recall the previously mentioned scaling factor that equals $I_o/(2V_T)$. It is inversely related to temperature as $V_T = kT/q$, where T is the absolute temperature in Kelvin (where k and q are the Boltzmann's constant and the electron charge, respectively). In the industrial temperature range of -40 to 120 °C, the scaling factor can deviate by almost ±30 % from its nominal value at room temperature. Consequently, it can be expected that the filter cutoff frequency will also suffer a deviation of ±30 %. This is illustrated in Figure 3-21, which shows the filter responses of the log-domain circuit operating at different temperatures. However, since we know that the frequency-scaling factor varies inversely with temperature, this problem can be easily remedied. One solution would be to make the current sources proportional to absolute temperature (PTAT) using a scheme similar to the one in band-gap references [8]. This will effectively ensure temperature-insensitivity for the resulting log-domain filter.

3.5 High-Order Filter Sensitivity-Comparative Studies

Doubly terminated lossless LC ladders that are designed for maximum power transfer are known to exhibit excellent low-sensitivity to the tolerances of its element values in the passband. It was pointed out that this advantageous property would be preserved with log-domain circuits [29]-[30]. To demonstrate this, we compare the Monte Carlo simulations (with 50 simulation runs) of two 7th-order lowpass Chebyshev log-domain filters based on the circuit of Figure 3-14(b), and a cascade of biquads [22] as described in Appendix C, respectively. Both filters feature 1 MHz cut-off frequency and 1 dB passband ripples with 100 μA bias current. Each capacitor is assigned a random variable having a Gaussian distribution with the mean value

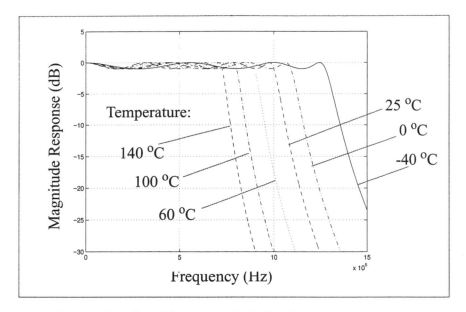

Figure 3-21: Cutoff frequency deviation due to temperature.

obtained from the previous synthesis procedure and a variance of 0.001 (i.e., 95% interval equals nominal value $\pm 6.2\%$). The results are plotted in Figure 3-22. As is evident, the ladder-based log-domain filter is far less sensitive to component variations. Although log-domain filtering belongs to the regime of nonlinear signal processing, the fact that LC ladder-based simulation has very good passband sensitivity properties is a testament that the log-domain circuit is capturing much of the LC ladder's internal operation.

To further verify the low passband sensitivity of the ladder-based log-domain filter, the 7th-order filter is compared to its cascade equivalence by assigning a random variable to each transistor area $(I_S)^\dagger$ with a variance of 0.001 and a mean as defined by the model statement. One hundred SPICE simulations are carried out. From the one hundred simulated frequency sweeps, the mean and the standard deviation (std) of the frequency response are computed (on a frequency-point by frequency-point basis). The final results are shown in Figure 3-23 Again, the ladder structure demonstrates a better sensitivity to area mismatches than the cascade structure.

---

†. The nonideal effects of transistor area mismatch will be analytically described in Chapter 5.

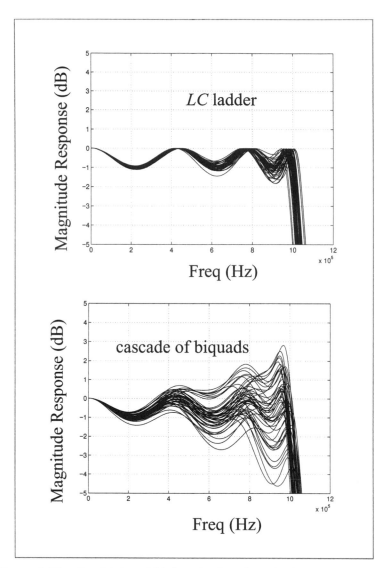

Figure 3-22: Sensitivity of high-order log-domain filters on capacitor variations.

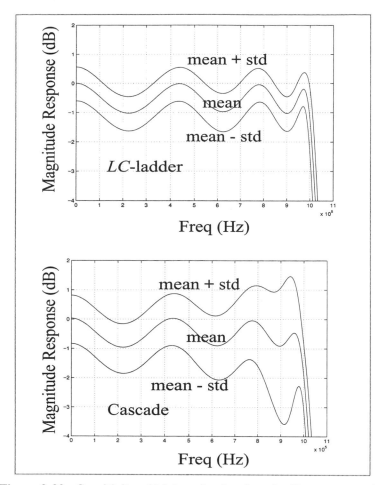

Figure 3-23: Sensitivity of high-order log-domain filters on transistor area mismatches.

3.6 Approximate Realization of Elliptic Filter Response

So far, only log-domain filters with Chebyshev characteristics are demonstrated. Here, we would apply the basic synthesis technique and investigate its feasibility to realize finite transmission zeros, or elliptical-type responses.

In order to limit the complexity of the derivation and the size of the ensuing circuit, the design will be targeted to a third-order elliptic filter. The filter characteristics are chosen as:

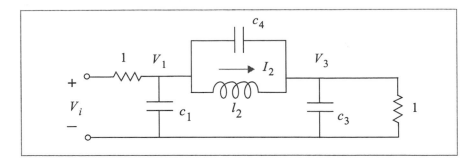

Figure 3-24: A 3rd-order elliptic lowpass *LC* ladder circuit.

Cutoff Frequency = 1 MHz

Max. Passband Ripple = 1 dB

Min. Stopband attenuation = 30 dB

From filter design table or CAD tools, the *LC* ladder that meets these specifications is given in Figure 3-24. The component values are summarized in Table 3-3.

By modified nodal analysis[†], the capacitor nodes of c_1 and c_3 are assigned voltage variables V_1 and V_3 respectively, while the inductor l_2 is given a current variable I_2. Again, the output is a scaled version of V_3 to account for the passband loss of the filter according to

$$V_{out} = 2V_3 \qquad (3.33)$$

The inter-relationships of these voltage and current variables can be written using straightforward circuit analysis as:

Name	Value	Name	Value
c_1	0.2647 µF	l_2	0.1197 µH
c_3	0.2647 µF	c_4	0.0480 µF

Table 3-3: Component values for the 3rd-order elliptic *LC* ladder.

---

†. The optimal selection of signal variables for a non-canonical *LC* ladder (where the number of reactive elements is larger than the filter order) will be discussed in Section 4.2.2 of the next chapter. It will be presented in the context of achieving a low-noise low-sensitivity state-space formulation.

$$V_1 = \frac{1}{c_1} \cdot \int \left\{ V_i - V_1 - I_2 - c_4 \frac{d}{dt}(V_1 - V_3) \right\} dt \qquad (3.34a)$$

$$I_2 = \frac{1}{l_2} \cdot \int (V_1 - V_3) dt \qquad (3.34b)$$

$$V_3 = \frac{1}{c_3} \cdot \int \left\{ I_2 - V_3 + c_4 \frac{d}{dt}(V_1 - V_3) \right\} dt \qquad (3.34c)$$

Further algebraic manipulation allows us to write the above equations as:

$$V_1 = \frac{1}{c_1 + c_4} \cdot \int \left\{ V_i - V_1 - I_2 + c_4 \frac{d}{dt} V_3 \right\} dt \qquad (3.35a)$$

$$I_2 = \frac{1}{l_2} \cdot \int (V_1 - V_3) dt \qquad (3.35b)$$

$$V_3 = \frac{1}{c_3 + c_4} \cdot \int \left\{ I_2 - V_3 + c_4 \frac{d}{dt} V_1 \right\} dt \qquad (3.35c)$$

Therefore, the SFG of Figure 3-25(a) can be drawn.

As should be expected from the similarity of their LC ladders, the SFG of the elliptic filter is very much like the one for an all-pole filter, e.g., Chebyshev filter. The only difference lies in the differentiators (circled in Figure 3-25(a)), which are caused by the series capacitor between nodes V_1 and V_3. As will be shown shortly, a single capacitor can be used to perform the differentiator operation in the log-domain filter as long as a simple approximation is made.

The next step in the design of the filter is to redraw the SFG in the log-domain as Figure 3-25(b) according to the rules discussed in Section 3.2. The differentiator sections can be left as is without changing the overall linear input-output relationship of the filter. Unfortunately, due to the lack of a log-domain differentiator, the circuit implementation is not readily available. However, we know that in other filter technologies such as active-RC, a simple capacitor is used to implement the differentiator. Let us, for the sake of argument, assume that a capacitor placed between nodes $\hat{V}_1$ and $\hat{V}_3$ of the Chebyshev filter will give the desired elliptic transfer function. This would result in the circuit shown in Figure 3-26.

In order to verify that the circuit of Figure 3-26 corresponds to the SFG of Figure 3-25(b), we write an expression for the current flowing into the capacitor C_3 (at node $\hat{V}_3$) due to the positive log-domain integrator and the floating capacitor C_4. The current through C_3 is thus given by

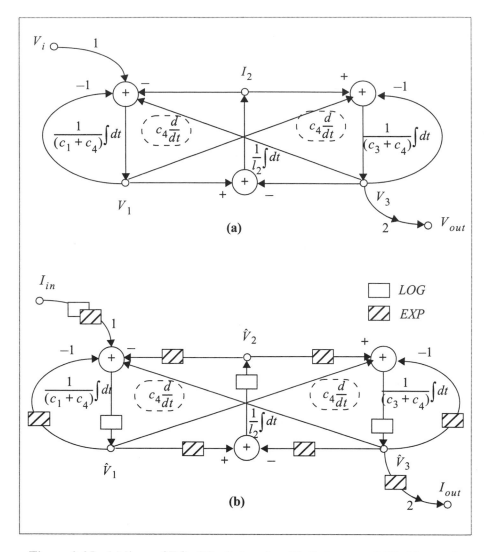

Figure 3-25: (a) linear SFG of the 3rd-order elliptic lowpass *LC* ladder, and (b) the equivalent log-domain SFG.

$$C_3\frac{d}{dt}\hat{V}_3 = C_4\frac{d}{dt}(\hat{V}_1 - \hat{V}_3) + I_o e^{\frac{\hat{V}_2 - \hat{V}_3}{2V_T}} - I_o \qquad (3.36)$$

Multiply through by $e^{\hat{V}_3/(2V_T)}$ gives

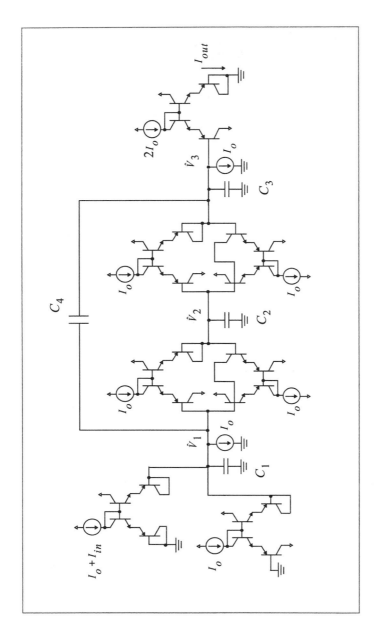

Figure 3-26: Circuit diagram of a 3rd-order elliptic lowpass log-domain filter.

$$C_3 \cdot e^{\frac{\hat{V}_3}{2V_T}} \cdot \frac{d}{dt}\hat{V}_3 = \left(C_4 \cdot e^{\frac{\hat{V}_3}{2V_T}} \cdot \frac{d}{dt}\hat{V}_1 \right) - \left(C_4 \cdot e^{\frac{\hat{V}_3}{2V_T}} \cdot \frac{d}{dt}\hat{V}_3 + I_o e^{\frac{\hat{V}_2}{2V_T}} - I_o e^{\frac{\hat{V}_3}{2V_T}} \right) \quad (3.37)$$

We must now make the following *approximation* to establish the correspondence between the elliptic filter circuit and the log-domain SFG:

$$e^{\frac{\hat{V}_3}{2V_T}} = e^{\frac{\hat{V}_1}{2V_T}} \qquad (3.38)$$

While such an assumption is not immediately obvious, practical experience with the log-domain elliptic filter has proven it to be a reasonable approximation [30]. Applying (3.38) to the first term of the right-hand side of (3.37), we have

$$C_3 \cdot e^{\frac{\hat{V}_3}{2V_T}} \cdot \frac{d}{dt}\hat{V}_3 = \left(C_4 \cdot e^{\frac{\hat{V}_1}{2V_T}} \cdot \frac{d}{dt}\hat{V}_1 \right) - \left(C_4 \cdot e^{\frac{\hat{V}_3}{2V_T}} \cdot \frac{d}{dt}\hat{V}_3 + I_o e^{\frac{\hat{V}_2}{2V_T}} - I_o e^{\frac{\hat{V}_3}{2V_T}} \right) \quad (3.39)$$

By means of chain-rule, we can write

$$\frac{2V_T C_3}{I_o} \cdot \frac{d}{dt}\left(I_o e^{\frac{\hat{V}_3}{2V_T}} - I_o \right) = \frac{2V_T C_4}{I_o} \cdot \frac{d}{dt}\left(I_o e^{\frac{\hat{V}_1}{2V_T}} - I_o \right) - \frac{2V_T C_4}{I_o} \cdot \frac{d}{dt}\left(I_o e^{\frac{\hat{V}_3}{2V_T}} - I_o \right)$$

$$+ \left(I_o e^{\frac{\hat{V}_2}{2V_T}} - I_o \right) - \left(I_o e^{\frac{\hat{V}_3}{2V_T}} - I_o \right) \qquad (3.40)$$

With the *EXP* notations as defined in (2.4), the last expression is simplified to be

$$EXP(\hat{V}_3) = \frac{I_o}{2V_T(C_3 + C_4)} \cdot \int \left\{ \frac{2V_T C_4}{I_o} \cdot \frac{d}{dt}[EXP(\hat{V}_1)] \right.$$

$$\left. + EXP(\hat{V}_2) - EXP(\hat{V}_3) \right\} dt \qquad (3.41)$$

A comparison between (3.41) and the log-domain SFG of Figure 3-25(b) shows that the floating capacitor will indeed implement the signal differentiation as required by an elliptic filter. Of course, this result is sound only if the approximation of (3.38) is valid. Similar conclusion can be drawn if the above exercise is repeated on the node $\hat{V}_1$, where a negative differentiation will be effectively realized as follows,

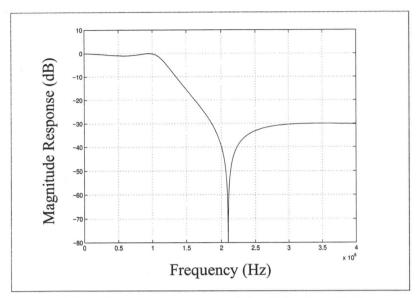

Figure 3-27: **Frequency response of the 3rd-order elliptic log-domain filter.**

$$EXP(\hat{V}_1) = \frac{I_o}{2V_T(C_1 + C_4)} \cdot \int \left\{ I_{in} - \frac{2V_T C_4}{I_o} \cdot \frac{d}{dt}[EXP(\hat{V}_3)] \right. \tag{3.42}$$

$$\left. - EXP(\hat{V}_1) - EXP(\hat{V}_2) \right\} dt$$

Finally, the following correspondence can be made between the log-domain filter (Figure 3-26) and its passive LC ladder prototype (Figure 3-24):

$$V_i \Leftrightarrow I_{in} \qquad V_1 \Leftrightarrow \hat{V}_1 \qquad I_2 \Leftrightarrow \hat{V}_2 \qquad V_3 \Leftrightarrow \hat{V}_3 \qquad V_{out} \Leftrightarrow I_{out}$$

$$\{C_1, C_2, C_3, C_4\} = \frac{I_o}{2V_T} \cdot \{c_1, l_2, c_3, c_4\} \tag{3.43}$$

Figure 3-27 shows the simulated frequency response of the log-domain circuit. Clearly, a near-perfect 3rd-order lowpass elliptic function is observed. Of course, the accuracy of this method hinges on the approximation depicted in (3.38), which can be questionable for high-order filter functions or large signal levels. Besides, some flexibility of the bias current tuning is lost. An exact synthesis scheme (with no approximation) is always preferred. Our next chapter on a state-space synthesis will address this issue.

3.7 Summary

In this chapter, the general method for the design of high-order log-domain filters based on the method of operational simulation of an *LC* ladder was described. This procedure can be understood as a straightforward extension of the conventional linear filter synthesis techniques. After the linear signal-flow graph is derived from the prototyping passive *LC* ladder, it will be made log-domain compatible by adding the *LOG* and *EXP* blocks according to a simple set of rules. Consequently, the log-domain integrators described in the previous chapter can be systematically applied to construct these filters. To illustrate this systematic synthesis method, four log-domain filter examples have been shown. The filter cutoff (or center) frequencies can be conveniently tuned by varying the bias level. By means of Monte-Carlo simulation, it is shown that the log-domain ladder filter exhibits very good passband sensitivity. This confirms the fact that the log-domain circuit is indeed capturing the low sensitivity properties of the *LC* ladder network.

To realize elliptic filter function following the same procedure, we encountered the problem related to the lack of a log-domain differentiator. We manage to overcome this by introducing an approximation, which suggests a floating capacitor to be connected across two log-domain voltage nodes. Although simulation results seem to support this method, this synthesis is not exact. It may result in higher distortion or excess deviations in the filter response.

Motivated by this observation, we will consider another synthesis method in the next chapter that is based on a state-space formulation. By doing so, we rely exclusively on integrators, thereby enabling the realization of arbitrary transfer functions, such as elliptic filter responses.

CHAPTER 4   Log-Domain Filter Synthesis-II: State-Space Formulation

As shown in Chapter 3, exact all-pole filter functions, such as those that realize a Butterworth or Chebyshev transfer function, can be synthesized using the method of operational simulation of LC ladder networks. The resulting filters feature easy tunability, excellent passband sensitivity and low noise properties. However, due to the lack of a log-domain differentiator, realizing finite transmission zeros exactly, such as those used in an elliptic filter function, is not possible. Therefore, a more versatile synthesis technique, based exclusively on integrators becomes necessary.

It is well known that arbitrary filter transfer functions can be synthesized from a system of equations in the form of a state-space representation. A state-space formulation consists of a set of first-order differential equations. As we shall discover in this chapter, it offers several benefits for the realization of high-order log-domain filters. There is a one-to-one correspondence between the mathematical formulation and the circuit realization. This facilitates systematic circuit implementation and makes debugging a simpler task. In addition, all of the capacitors in a current-mode implementation of a state-space filter have one terminal connected to ground. This makes the filter more amenable for IC implementation. Finally, a state-space formulation provides a very general solution for realizing arbitrary transfer functions.

In this chapter, we are going to demonstrate how a state-space representation of an Nth-order transfer function can be used to realize arbitrary high-order high-performance log-domain filters.

4.1   Frey's Exponential State-Space Synthesis

Research into log-domain filters was rekindled by Frey in 1993 using an "exponential state-space" synthesis method [22]. Its general idea is as follows. Consider a dynamic state-space representation of an arbitrary filter function:

$$\frac{dx(t)}{dt} = Ax(t) + bu(t)$$

$$y(t) = c^T x(t) + du(t)$$
(4.1)

where $x(t) = (x_1(t), x_2(t), ..., x_n(t))^T$ is the vector of state variables, $u(t)$ and $y(t)$ are the scalar input and output signals, respectively, and A is a $N \times N$ matrix, while b and c are $N \times 1$ vectors. The "d" term denotes a scaled feed-through from the filter input to the output. Also, if we map the state variables $x_i(t)$ and scalar input $u(t)$ to the exponential of voltages $V_i(t)$ and $U(t)$ according to[†]

$$x_i(t) = e^{(V_i(t))/V_T}$$

$$u(t) = I_o e^{(U(t))/V_T}$$
(4.2)

then (4.1) can be rewritten as

$$\frac{1}{V_T} \cdot \left(\dot{V}_1 e^{\frac{V_1}{V_T}}, \dot{V}_2 e^{\frac{V_2}{V_T}}, ..., \dot{V}_n e^{\frac{V_n}{V_T}} \right)^T = A \cdot \left(e^{\frac{V_1}{V_T}}, e^{\frac{V_2}{V_T}}, ..., e^{\frac{V_n}{V_T}} \right)^T + b \cdot I_o e^{\frac{U}{V_T}}$$

$$y = c^T \cdot \left(e^{\frac{V_1}{V_T}}, e^{\frac{V_2}{V_T}}, ..., e^{\frac{V_n}{V_T}} \right)^T + D \cdot I_o e^{\frac{U}{V_T}}$$
(4.3)

where the time reference has been dropped from the expression to simplify its presentation. Multiply the first expression in (4.3) by $(C_i V_T) \exp(-V_i / V_T)$, we obtain

$$C_i \dot{V}_i = \sum_{j=1}^{n} \left(C_i V_T A_{ij} e^{\frac{V_j - V_i}{V_T}} + C_i V_T b_i I_o e^{\frac{U - V_i}{V_T}} \right)$$
(4.4)

[†]. Notice that our convention of specifying log-domain signals by circumflex (^) is not followed here, in order to present the original look of the exponential state-space synthesis method in [22].

where the C_i are arbitrary constants. In order to interpret (4.4) as a set of KCL equations to be realized by actual circuit elements, we will rewrite it as

$$C_i \dot{V}_i = \sum_{j=1}^{n} I_S e^{\frac{V_j + V_{aij} - V_i}{V_T}} + I_S e^{\frac{U + V_{bi} - V_i}{V_T}}$$

$$\text{where} \qquad V_{aij} = V_T \cdot \ln\left(\frac{C_i A_{ij} V_T}{I_S}\right) \tag{4.5}$$

$$\text{and} \qquad V_{bi} = V_T \cdot \ln\left(\frac{C_i b_i V_T I_o}{I_S}\right)$$

For the circuit implementation, the following observations are utilized:

1. The term $C_i \dot{V}_i$ on the left side of (4.5) represents the current flowing into the grounded capacitor C_i tied to node i.

2. Each item on the right side of (4.5) denotes a current flowing through a diode, with a "composite" voltage of, say, $V_j + V_{aij} - V_i$ or $U + V_{bi} - V_i$ across it.

3. V_{aij} represents a diode voltage due to a forward-biased current of magnitude $\left| C_i A_{ij} V_T \right|$. Similar argument also holds for V_{bi}.

4. When bipolar transistors are used, $V_j + V_{aij}$ corresponds to the jth node voltage (V_j) being level-shifted by a diode-connected transistor ($Q1$) with base-emitter voltage equals V_{aij}. This voltage is then applied to the base of another bipolar transistor ($Q2$) whose emitter is in turn connected to the ith node voltage V_i. The collector current of $Q2$ will then implement the desired items on the right side of (4.5). This is shown in Figure 4-1.

   The above procedure is repeated for all state-space equations. Similar manipulations also apply to the second equation in (4.3), which is the input/output relation. In summary, the exponential state-space synthesis method involves transforming the state-space variables into node voltages by an appropriate exponential mapping, so that bipolar transistors can be used to realize the resulting expressions.

   So far, to the best of our knowledge, only biquadratic filters have been designed from this approach. Strictly speaking, although high-order log-domain filters are possible by cascading biquads [22], high-order synthesis from a single state-space model has never been successfully tried. This fact exposes a major weakness of this approach: the complexity of the state-space equations can easily

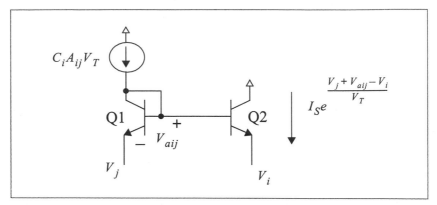

Figure 4-1: Simplified demonstration of exponential state-space synthesis method in [22].

become unmanageable as the filter order increases, despite the fact that the synthesis is theoretically sound regardless of the filter order. Besides, due to its purely mathematical manner, not much physical insight is offered. Practically, the synthesis requires a fair amount of ad-hoc circuit tricks, thus making the design procedure rather unsystematic and limited to log-domain experts. Therefore, we are going to derive an alternative state-space synthesis methodology that is related to, but more systematic than, the above procedure. Based on the log-domain integrator sub-circuits, high-order log-domain filters can then be synthesized from any arbitrary state-space formulations in a straightforward manner.

4.2   Towards a State-Space Log-Domain Filter Realization

Before we embark on the synthesis of log-domain filters, it would be worthwhile to spend some time to reflect on the physical meanings of the state-space equations. By probing beyond their simple mathematical form, we will achieve a more intuitive understanding of what the equations represent. Afterwards, a table of log-domain functional blocks will be provided. As such, synthesis of state-space log-domain filters will become a straightforward exercise of mapping the state-space equations into various log-domain circuits (namely the log-domain integrators). This modular synthesis scheme will greatly reduce the complexity involved, especially in the case of synthesizing high-order filters.

4.2.1   State-Space Formulation Revisited

For ease of discussion, the Nth-order state-space system described by (4.1) is recaptured below,

$$\frac{dx(t)}{dt} = Ax(t) + bu(t)$$

$$y(t) = c^T x(t) + du(t)$$

(4.6)

in which the time-dependency of the signals is explicitly shown. Again, $u(t)$ and $y(t)$ denote the input and output signals of the overall filter system and x is a vector of state-variables. Parameters A, b, c and d are the linear coefficients relating these input, output and state-variables.

The Laplace transformation of (4.6) is given by

$$sX(s) = AX(s) + bU(s)$$

$$Y(s) = c^T X(s) + dU(s)$$

(4.7)

where $X(s)$, $Y(s)$, and $U(s)$ represent the Laplace transforms of $x(t)$, $y(t)$ and $u(t)$, respectively. Eliminating the state variable vector $X(s)$ from (4.7), the transfer function realized by the above system of equations is given by

$$T(s) = \frac{Y(s)}{U(s)} = c^T (sI - A)^{-1} b + d$$

(4.8)

The poles of the system described by (4.8) are given by the eigenvalues of A, while the system zeros are related to all four state-space parameters.

Notice that since $T(s)$ is a scalar function, an identical system can be realized from its transpose. From (4.8) we can write

$$[T(s)]^T = [c^T (sI - A)^{-1} b + d]^T$$

(4.9)

Seeing that $[T(s)]^T = T(s)$, an alternative representation can be written as

$$T(s) = b^T (sI - A^T)^{-1} c + d$$

(4.10)

By comparing (4.8) to (4.10), the relationships between the original coefficients and that of the transposed system are found to be:

$$A \Leftrightarrow A^T \qquad b \Leftrightarrow c \qquad d \Leftrightarrow d$$

(4.11)

These transpose system relationships will prove useful for the implementation of certain transfer functions, as will be demonstrated in Section 4.3.2.

One usually begins a design by working with a frequency normalized state-space formulation of the desired transfer function, largely for numerical accuracy reasons. This typically implies that the upper edge of the passband region of the filter response is set to 1 rad/s. To frequency denormalize the state-space formulation such that the passband edge corresponds to a new frequency, say f_p expressed in Hz, we make use of the frequency translation, $s \rightarrow s/(2\pi f_p)$, and substitute this into (4.7) according to

$$\frac{s}{2\pi f_p} X(s) = AX(s) + bU(s)$$
$$Y(s) = c^T X(s) + dU(s)$$

(4.12)

Rearranging such that the state-space formulation takes on its usual form, we write

$$sX(s) = 2\pi f_p AX(s) + 2\pi f_p bU(s)$$
$$Y(s) = c^T X(s) + dU(s)$$

(4.13)

Frequency denormalization is therefore achieved by multiplying the A and b matrices by the desired frequency scaling factor $2\pi f_p$.

For a particular $T(s)$, there exist infinite sets of state-space parameters (A, b, c, d) that can satisfy (4.8). This is a direct result of the similarity transformation [72]. Consider a linear transformation of the state-variables written as

$$X = TX'$$

(4.14)

where T is an invertible $N \times N$ matrix. Substituting (4.14) into (4.7), we can write

$$sTX'(s) = ATX'(s) + bU(s)$$
$$Y(s) = c^T TX'(s) + dU(s)$$

(4.15)

Multiplying both sides (from the left) by T^{-1}, we get

$$sX'(s) = T^{-1}ATX'(s) + T^{-1}bU(s)$$
$$Y(s) = (T^T c)^T X'(s) + dU(s)$$

(4.16)

where we see

$$(A, \ b, \ c, \ d) \rightarrow (T^{-1}AT, \ T^{-1}b, \ T^T c, \ d)$$

(4.17)

If we substitute the state-space parameters of (4.17) into (4.8), we can write the system transfer function as

$$T(s) = (T^T c)^T (sI - T^{-1} A T)^{-1} (T^{-1} b) + d$$
$$= c^T (sI - A)^{-1} b + d \qquad (4.18)$$

which, after further algebraic manipulations, we return to (4.8). We leave the details to the reader to confirm. This result suggests that for any invertible matrix T, a different set of state-space parameters can provide the same circuit transfer function.

Under ideal conditions, perfect and identical filter responses will always result regardless of our choice of parameters. Unfortunately, this is hardly true in practice. Different formulations will result in circuits that exhibit various levels of deviations. It is then crucial for the designer to find a specific set of equations that will yield the best filter realization, such as one with small coefficient (component) spread, low sensitivity and good noise properties. A simple and efficient method will be demonstrated in the next section.

4.2.2 Design of Low-Noise Low-Sensitivity Filters

A detailed analysis leading to the selection of an optimal set of state-space parameters is provided by Snelgrove and Sedra [54]. It involves a set of intermediate functions that provide a link between the state-space coefficients, and the noise and sensitivity properties of the resulting state-space system. In this way, the performance space can be searched first without regard to the topology or actual component values. Hence, a more efficient search for a state-space structure with the lowest noise and sensitivity behavior is achieved. Such an analysis has led to a direct procedure for selecting the state-space coefficients for a low-noise low-sensitivity filter. It is based on selecting the appropriate intermediate functions from the LC ladder network [55]-[56]. The procedure for obtaining the state-space coefficients from an arbitrary LC ladder is presented in [57]. The steps can be best described through an example.

Consider again the 3rd-order low-pass elliptic filter as discussed in Section 3.6. An optimal set of state-space coefficients will be derived from the LC ladder prototype of Figure 3-24. This ladder circuit is recaptured in Figure 4-2, while the component values are summarized in Table 4-1 for a normalized passband frequency of 1 rad/s.

We shall now relate physical quantities of the LC ladder (such as voltages, currents, RLC components) to the state-space formulation variables and parameters: X, A, b, c and d.

From the LC ladder, a total of N voltages and currents will be chosen to represent the N state variables x. In the case of a canonic ladder, where the number of reactive elements, i.e., capacitors and inductors, equals N, we simply choose all of the inductor currents and capacitor voltages. However, for a non-canonic ladder, where $M > N$ (M being the number of reactive elements) $M - N$ states need to be eliminated. As such, the remaining N states will form the N linearly independent

Name	Value	Name	Value
R_S	1 Ω	C_2	0.30180 F
R_L	1 Ω	C_3	1.66332 F
C_1	1.66332 F	L_2	0.75206 H

Table 4-1: Component values for the 3rd-order elliptic *LC* ladder

equations. In general, extra states are eliminated by removing a capacitor voltage from each cutset (capacitors that are connected in a loop), or an inductor current from each tieset (inductors that connect to a common node). Evidently, there are many possible ways to remove these $M - N$ states. However, it has been shown in [57] that good filter realizations can be obtained by removing the cutset or tieset states that are part of resonant tanks.

For the *LC* ladder of Figure 4-2, the capacitor voltage V_{C2} represents a state that is part of a cutset and a resonant tank. According to the rule discussed above, it is removed from our set of state-variables. Therefore, for the 3rd-order filter system $(N = 3)$, the three states are left to be V_{C1}, I_{L2} and V_{C3}. Applying KCL on the voltage nodes of V_{C1} and V_{C3}, the following two state equations are derived:

$$sC_1V_{C1} + sC_2(V_{C1} - V_{C3}) + I_{L2} + \frac{V_{C1} - V_{in}}{R_S} = 0 \qquad (4.19)$$

$$sC_3V_{C3} - sC_2(V_{C1} - V_{C3}) - I_{L2} + \frac{V_{C3}}{R_L} = 0 \qquad (4.20)$$

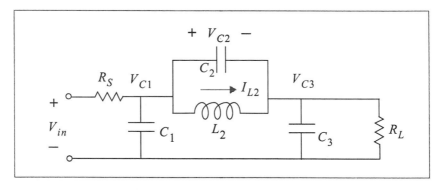

Figure 4-2: A 3rd-order *LC* ladder circuit.

Alternatively, if we apply KVL around the C_1-L_2-C_3 loop, the third state equation is found,

$$sL_2I_{L2} - V_{C1} + V_{C3} = 0 \qquad (4.21)$$

To achieve the same transfer function as implemented by the LC ladder, we select

$$V_{out} = V_{C3} \qquad (4.22)$$

However, for a 0-dB passband gain, it is shown in [57] that the following condition should be satisfied,

$$V_{out} = 2\sqrt{\frac{R_L}{R_S}} \cdot V_{C3} \qquad (4.23)$$

The system of state equations (4.19)-(4.23) can be combined and rewritten using matrix notation as below:

$$s\begin{bmatrix} (C_1 + C_2) & 0 & -C_2 \\ 0 & L_2 & 0 \\ -C_2 & 0 & (C_2 + C_3) \end{bmatrix}\begin{bmatrix} V_{C1} \\ I_{L2} \\ V_{C3} \end{bmatrix} + \begin{bmatrix} \left(\frac{1}{R_S}\right) & 1 & 0 \\ -1 & 0 & 1 \\ 0 & -1 & \left(\frac{1}{R_L}\right) \end{bmatrix}\begin{bmatrix} V_{C1} \\ I_{L2} \\ V_{C3} \end{bmatrix} = \begin{bmatrix} \left(\frac{1}{R_S}\right) \\ 0 \\ 0 \end{bmatrix}V_{in}$$

$$(4.24)$$

$$V_{out} = \begin{bmatrix} 0 & 0 & \left(2\sqrt{\frac{R_S}{R_L}}\right) \end{bmatrix}\begin{bmatrix} V_{C1} \\ I_{L2} \\ V_{C3} \end{bmatrix} + 0$$

Recognizing that the second equation of (4.24) bears identical form to the second equation of (4.7), the state-space parameters c^T and d are immediately found:

$$c^T = \begin{bmatrix} 0 & 0 & \left(2\sqrt{\frac{R_S}{R_L}}\right) \end{bmatrix}; \qquad d = 0 \qquad (4.25)$$

On the other hand, to determine the A and b matrices, a simple manipulation on the first equation of (4.24) is needed. It can be rewritten in the form,

$$[sC + G]X = WV_{in} \qquad (4.26)$$

where the matrices C, G, and W are respectively given by

$$
C = \begin{bmatrix} (C_1 + C_2) & 0 & -C_2 \\ 0 & L_2 & 0 \\ -C_2 & 0 & (C_2 + C_3) \end{bmatrix}; \quad G = \begin{bmatrix} \left(\dfrac{1}{R_S}\right) & 1 & 0 \\ -1 & 0 & 1 \\ 0 & -1 & \left(\dfrac{1}{R_L}\right) \end{bmatrix};
$$

$$
W = \begin{bmatrix} \left(\dfrac{1}{R_S}\right) \\ 0 \\ 0 \end{bmatrix}
$$

(4.27)

Given that the inverse of the C matrix exists, (4.26) can be rearranged as

$$
sX = -C^{-1}GX + C^{-1}WV_{in}
$$

(4.28)

Comparing (4.28) with the first equation of the state-space formulation (4.7), we can now identify the state space parameters A and b to be:

$$
A = -C^{-1}G; \qquad b = C^{-1}W
$$

(4.29)

Equations (4.25) and (4.29) effectively relate the A, b, c, d parameters to the physical quantities described by the LC ladder circuit. Their numerical values can be computed by the straightforward substitution shown below. Applying the LC ladder component values of Table 4-1 into (4.27), the C, G and W matrices are respectively found to be:

$$
C = \begin{bmatrix} 1.96513 & 0 & -0.30180 \\ 0 & 0.75206 & 0 \\ -0.30180 & 0 & 1.96513 \end{bmatrix}; \quad G = \begin{bmatrix} 1 & 1 & 0 \\ -1 & 0 & 1 \\ 0 & -1 & 1 \end{bmatrix};
$$

$$
W = \begin{bmatrix} 1 \\ 0 \\ 0 \end{bmatrix}
$$

(4.30)

Substitute the results of (4.30) into (4.29), and perform the matrix mathematics, the A and b state-space parameters are obtained as:

$$
A = \begin{bmatrix} -0.52116 & -0.44113 & -0.08004 \\ 1.32968 & 0 & -1.32968 \\ -0.08004 & 0.44113 & 0.52116 \end{bmatrix}; \quad b = \begin{bmatrix} 0.52116 \\ 0 \\ 0.08004 \end{bmatrix}
$$

(4.31)

On the other hand, the c^T and d parameters are simply given by,

$$c^T = \begin{bmatrix} 0 & 0 & 2 \end{bmatrix}; \qquad d = 0 \qquad (4.32)$$

after substituting the values of R_L and R_S (from Table 4-1) into (4.25). Now, we have achieved the state-space formulation for a third-order elliptic lowpass filter. In summary, it involves the steps of:

1. Finding an LC ladder which meets the desired filter specification

2. Select a total of N capacitor voltages and inductor currents as state variables

3. Write their inter-relationships by straightforward circuit analysis

4. Map the resulting system of state equations into A, b, c and d parameters.

Notice that the procedures demonstrated for our previous example are applicable to any arbitrary LC network. In addition, as the state-space formulation is derived directly from the LC ladder, this realization will have very good sensitivity and noise properties. For instance, an aggregate measure of the overall sensitivity function, $Sen(s)$, is often written as

$$Sen(s) = \sqrt{\sum_{ij} \left|S_{A_{ij}}^{T(s)}\right|^2 + \sum_i \left|S_{b_i}^{T(s)}\right|^2 + \sum_i \left|S_{c_i}^{T(s)}\right|^2 + \left|S_d^{T(s)}\right|^2} \qquad (4.33)$$

where the symbol $S_x^{T(s)}$ denotes the sensitivity function so that the fractional deviation of the transfer function $T(s)$ is related to an arbitrarily small Δx by

$$\frac{\Delta T}{T} \approx S_x^T \cdot \left(\frac{\Delta x}{x}\right) \qquad (4.34)$$

Evaluating the sensitivity expression (4.33) for our example in Figure 4-2 (i.e, the equations of (4.31) and (4.32)) leads to $|Sen(s)| \sim 10$ over the passband region. This amount of sensitivity is similar to that obtained for the corresponding LC ladder network of Figure 4-2 with a sensitivity function having the form

$$Sen(s) = \sqrt{\sum_i \left|S_{L_i}^{T(s)}\right|^2 + \sum_i \left|S_{C_i}^{T(s)}\right|^2 + \sum_i \left|S_{R_i}^{T(s)}\right|^2}. \qquad (4.35)$$

The final step in selecting the state-space parameters is to scale the system for maximum dynamic range. This is easily achieved through the application of the similarity transform described in Section 4.2.1. Following the dynamic range scaling rule of Chapter 3, we first find the spectral peaks associated with each state-variable by sweeping the input signal over a desired frequency band. This can be achieved by simulating the LC ladders in SPICE, or by evaluating the following state-variable equation derived from (4.7)

$$X(s) = [sI - A]^{-1} \cdot b \tag{4.36}$$

over the frequency range of interest with $U(s) = 1$.

As denoted previously, the spectral peak value of the i-th state-variable is labeled as P_i. Next, a $N \times N$ diagonal matrix T is defined in terms of these spectral peaks as follows

$$T = \frac{1}{P_{max}} \begin{bmatrix} P_1 & 0 & \cdots & 0 & 0 \\ 0 & P_2 & \cdots & 0 & 0 \\ 0 & 0 & \cdots & 0 & 0 \\ 0 & 0 & \cdots & P_{N-1} & 0 \\ 0 & 0 & \cdots & 0 & P_N \end{bmatrix} \tag{4.37}$$

where P_{max} is to be the new spectral peak value of each state-variable. Subsequently, the parameters of the modified state-space formulations (denoted by the prime symbol) are given by (4.17) and repeated below:

$$A' = T^{-1}AT \qquad b' = T^{-1}b \qquad c' = T^T c \qquad d' = d \tag{4.38}$$

For the above running example given in (4.31), we find through simulation that the spectral peak values are 0.6647, 1.2420 and 0.5. To scale the state-variable such that they all have equal peak values of unity, i.e. $P_{max}=1$, we establish the following T matrix:

$$T = \begin{bmatrix} 0.6647 & 0 & 0 \\ 0 & 1.2420 & 0 \\ 0 & 0 & 0.5000 \end{bmatrix} \tag{4.39}$$

Hence, using (4.38), we find the parameters for the dynamically scaled state-space system as

$$A' = \begin{bmatrix} -0.5212 & -0.8242 & -0.0602 \\ 0.7116 & 0 & -0.5353 \\ -0.1064 & 1.0957 & -0.5212 \end{bmatrix} ; \qquad b' = \begin{bmatrix} 0.7841 \\ 0 \\ 0.1601 \end{bmatrix} \tag{4.40}$$

and

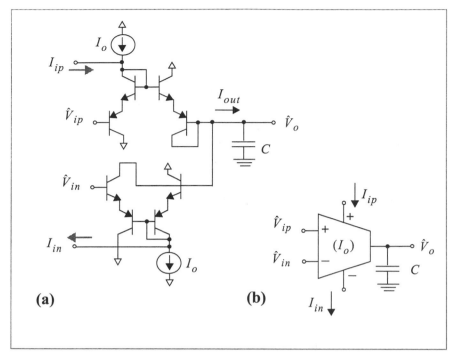

Figure 4-3: (a) The multiple-input log-domain integrator, and (b) its symbol.

$$c'^{T} = \begin{bmatrix} 0 & 0 & 1 \end{bmatrix}; \qquad d' = 0 \qquad\qquad (4.41)$$

4.2.3 A Collection of Log-Domain Functional Blocks

The set of state-space formulations can be realized by log-domain integrators and the corresponding *LOG* and *EXP* circuits systematically. To facilitate the discussion, we will first introduce a universal symbol for the log-domain building block as shown in Figure 4-3. It can be recognized as the simple multiple-input log-domain integrator (such as the one shown in Figure 2-2 of chapter 2) with the addition of two current signals, I_{ip} and I_{in}, superimposed on the dc bias currents [31], [73]. Similar to (2.2), its nodal equation written at the output terminal is given by

$$I_{out} = C \cdot \frac{d\hat{V}_o}{dt} = (I_o + I_{ip})e^{\frac{\hat{V}_{ip} - \hat{V}_o}{2V_T}} - (I_o + I_{in})e^{\frac{\hat{V}_{in} - \hat{V}_o}{2V_T}} \qquad (4.42)$$

Following our signal convention so far I_o denotes the dc bias current which is largely responsible for tuning the integrator time constant. By rearranging the inputs to the voltage and current ports, this cell is capable of realizing a variety of signal-processing functions in the log-domain. We have already seen some of its variations in Sections 2.1 and 2.2, such as the damped log-domain integrator and the input LOG operator. For ease of reference, a more thorough summary is produced in Table 4-2. To familiarize oneself with this array of functional blocks, readers are encouraged to verify each I_{out} expression listed in Table 4-2 based on (4.42), and the resulting transfer functions following the steps involved in deriving (2.2) to (2.5).

For simplicity, when the input currents, i.e. I_{ip} and I_{in}, both equal zero, the upper and lower current ports of the symbol are omitted. This is shown in cases (i) to (iv), (vii) and (viii) of Table 4-2. Besides, recall the damped log-domain integrator discussed in Section 2.1.2. When $\hat{V}_o = \hat{V}_{in}$, half of the integrator circuit can be simplified and equivalently replaced by a dc current source. Therefore, for cases (iii), (iv), (vii) and (viii), where the output node is tied to either one of the input voltage terminals, it is understood that the circuit of Figure 4-3 is simplified in the same manner.

Before we leave this section, we would like to point out that the log-domain integrator circuit of Figure 4-3(a) is only selected as an example to illustrate the basic idea. The log-domain state-space filters can be constructed using other more sophisticated log-domain integrator circuits, such as those described in Chapter 2. The derivations to be shown shortly will be equally applicable after a minor scaling factor is accounted for (see the footnote on page 56).

4.3   State-Space Synthesis of Log-Domain Filters

Using the state-space formulation and the log-domain integrator cells introduced in the previous section, we can systematically realize an arbitrary-order state-space filter [32]-[33]. The general synthesis example of an Nth-order system is first presented. We will see that the resulting log-domain circuit will exhibit direct correspondence to its mathematical description on a term by term basis. The theory will be applied to the synthesis of several log-domain filters. Continuing the 3rd-order elliptic lowpass state-space equations derived in (4.31)-(4.32), the corresponding log-domain circuit realization will be presented. To demonstrate the generality of this scheme, further circuit examples will also be demonstrated, which include a 3rd-order highpass filter and a 6th-order bandpass filter. Simulation results will be shown to verify the circuits.

4.3.1   A General N-th Order Log-Domain State-Space Filter

By applying a simple mapping to the linear state-space equations, the corresponding log-domain circuit realizations can be obtained. Toward that end, we will re-write the first equation of (4.6) in matrix form as

Log-Domain Building Blocks		Log-Domain Equations	Symbols
i	Positive Integrator	$$I_{out} = I_o e^{\frac{\hat{V}_{ip} - \hat{V}_o}{2V_T}} - I_o e^{\frac{-\hat{V}_o}{2V_T}}$$ $$EXP(\hat{V}_o) = \frac{I_o}{2V_T C} \cdot \int EXP(\hat{V}_{ip})dt$$	OTA (I_o) with $\hat{V}_{ip}$ at $+$, 0 at $-$; output $\hat{V}_o$, C, I_{out}
ii	Negative Integrator	$$I_{out} = I_o e^{\frac{-\hat{V}_o}{2V_T}} - I_o e^{\frac{\hat{V}_{in} - \hat{V}_o}{2V_T}}$$ $$EXP(\hat{V}_o) = \frac{-I_o}{2V_T C} \cdot \int EXP(\hat{V}_{in})dt$$	OTA (I_o) with 0 at $+$, $\hat{V}_{in}$ at $-$; output $\hat{V}_o$, C, I_{out}
iii	Positive Damped Integrator	$$I_{out} = I_o e^{\frac{\hat{V}_{ip} - \hat{V}_o}{2V_T}} - I_o$$ $$EXP(\hat{V}_o) = \frac{I_o}{2V_T C} \cdot \int [EXP(\hat{V}_{ip}) - EXP(\hat{V}_o)]dt$$	OTA (I_o) with $\hat{V}_{ip}$ at $+$, $\hat{V}_o$ at $-$; output $\hat{V}_o$, C, I_{out}
iv	Negative Damped Integrator	$$I_{out} = I_o - I_o e^{\frac{\hat{V}_{in} - \hat{V}_o}{2V_T}}$$ $$EXP(\hat{V}_o) = \frac{-I_o}{2V_T C} \cdot \int [EXP(\hat{V}_{in}) - EXP(\hat{V}_o)]dt$$	OTA (I_o) with $\hat{V}_o$ at $+$, $\hat{V}_{in}$ at $-$; output $\hat{V}_o$, C, I_{out}

Table 4-2: A list of log-domain building blocks (*to be continued on next page*).

Log-Domain Building Blocks		Log-Domain Equations	Symbols
v	Positive Integrator and Input *LOG*	$I_{out} = I_{ip} e^{\frac{-\hat{V}_o}{2V_T}}$ $EXP(\hat{V}_o) = \frac{I_o}{2V_T C} \cdot \int I_{ip}\, dt$	
vi	Negative Integrator and Input *LOG*	$I_{out} = -I_{in} e^{\frac{-\hat{V}_o}{2V_T}}$ $EXP(\hat{V}_o) = \frac{-I_o}{2V_T C} \cdot \int I_{in}\, dt$	
vii	Positive Output *EXP*	$I_{out} = I_o e^{\frac{\hat{V}_{ip}}{2V_T}} - I_o$	
viii	Negative Output *EXP*	$I_{out} = I_o - I_o e^{\frac{\hat{V}_{in}}{2V_T}}$	

Table 4-2: *(continued)* A list of log–domain building blocks

$$\frac{d}{dt}\begin{bmatrix} x_1(t) \\ x_2(t) \\ \vdots \\ x_N(t) \end{bmatrix} = \begin{bmatrix} A_{11} & A_{12} & \cdots & A_{1N} \\ A_{21} & A_{22} & \cdots & A_{2N} \\ \vdots & \vdots & \cdots & \vdots \\ A_{N1} & A_{N2} & \cdots & A_{NN} \end{bmatrix}\begin{bmatrix} x_1(t) \\ x_2(t) \\ \vdots \\ x_N(t) \end{bmatrix} + \begin{bmatrix} b_1 \\ b_2 \\ \vdots \\ b_N \end{bmatrix} u(t) \qquad (4.43)$$

which can be interpreted as N linear row equations. Each of these N row equations describes an integration (inverse of the differentiation operation) over the N state-variables $\{x_1(t), x_2(t), ..., x_N(t)\}$ and the input $u(t)$, weighted by the scalar factors defined in the A and b matrices. For instance, the first row of (4.43) can be written as

$$\frac{d}{dt}x_1(t) = A_{11}x_1(t) + A_{12}x_2(t) + ... + A_{1N}x_N(t) + b_1 u(t) \qquad (4.44)$$

or in integral form as

$$x_1(t) = \int (b_1 u(t) + A_{11}x_1(t) + A_{12}x_2(t) + ... + A_{1N}x_N(t))dt \qquad (4.45)$$

Here, we have seen that the state-variable $x_1(t)$ equals a weighted sum of the integrals of the input, itself (damping) and all other N-1 state variables. The input term has been moved from the last to the first operand of the integration function for future consideration. The scaling factors, as given by b_1, A_{11} to A_{1N}, are closely related to integrator time-constant, as we will see shortly.

Following the procedure of Chapter 3, we can now draw the SFG based on (4.45) as shown in Figure 4.4(a). Adding the *LOG* and *EXP* cells to the linear SFG in the appropriate manner, the log-domain SFG is obtained as shown in Figure 4.4(b). The variables in the two SFGs then are assumed to correspond according to the following mappings:

$$u \Leftrightarrow I_{in} \qquad x_1 \Leftrightarrow \hat{V}_1 \qquad x_2 \Leftrightarrow \hat{V}_2 \qquad ... \qquad x_N \Leftrightarrow \hat{V}_N \qquad (4.46)$$

Finally, the log-domain filter circuit is obtained by replacing the appropriate branches of the log-domain SFG with multiple-input log-domain integrators and the input *LOG* circuit. The result is illustrated in Figure 4-5 using the universal log-domain cell symbol to represent each branch of the SFG. From Table 4-2, the first block in Figure 4-5 can be recognized as the input *LOG* operator[†] (and a log-domain integrator), the second block a log-domain negative damped integrator, while the rest are positive log-domain integrators. Employing the I_{out} expressions documented in Table 4-2, the following KCL equation can be written at the capacitive node ($\hat{V}_1$) by inspection as,

$$C\frac{d}{dt}\hat{V}_1 = I_{in}e^{\frac{-\hat{V}_1}{2V_T}} + \left[I_{A_{11}} - I_{A_{11}}e^{\frac{-\hat{V}_1}{2V_T}}\right]$$

$$+ \left[I_{A_{12}}e^{\frac{\hat{V}_2-\hat{V}_1}{2V_T}} - I_{A_{12}}\right] + \dots + \left[I_{A_{1N}}e^{\frac{\hat{V}_N-\hat{V}_1}{2V_T}} - I_{A_{1N}}\right]$$

$$(4.47)$$

Multiply both sides by $e^{\hat{V}_1/(2V_T)}$ and re-arrange, we have

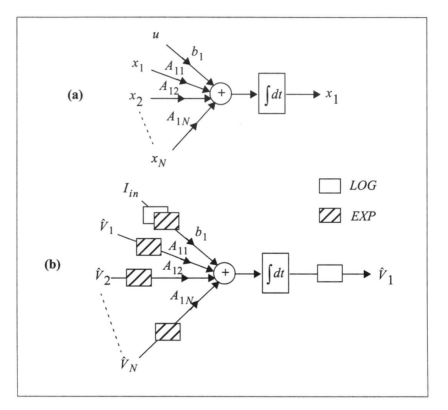

Figure 4-4: SFG for first-row of state-space integral equations: (a) linear SFG, and (b) equivalent log-domain SFG.

---

†. From (4.47), it can be seen that the bias current of this input *LOG* operator (i.e., the top cell in Figure 4-5) does not show up in the KCL equation where $\hat{V}_1$ is computed. This exposes the fact that the selection of this bias current can be arbitrary. Here, for simplicity, a current of I_o is chosen, which is the same current in realizing the *LOG*/*EXP* complementary mapping.

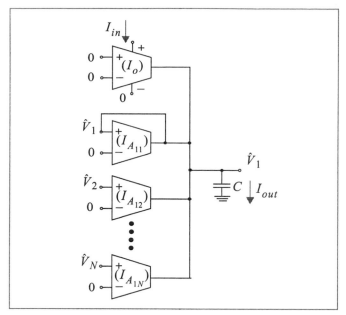

Figure 4-5: Equivalent log-domain circuit realizing the SFG shown in Figure 4.4 (b).

$$\frac{2V_TC}{I_o} \cdot \frac{d}{dt}\left[I_o e^{\frac{\hat{V}_1}{2V_T}} - I_o\right] = I_{in} + \frac{I_{A11}}{I_o} \cdot \left[I_o e^{\frac{\hat{V}_1}{2V_T}} - I_o\right]$$

$$+ \frac{I_{A12}}{I_o} \cdot \left[I_o e^{\frac{\hat{V}_2}{2V_T}} - I_o\right] + ... + \frac{I_{A1N}}{I_o} \cdot \left[I_o e^{\frac{\hat{V}_N}{2V_T}} - I_o\right] \qquad (4.48)$$

Making use of the familiar *LOG* and *EXP* mapping functions,

$$LOG(x) = 2V_T \ln\left(\frac{I_o + x}{I_o}\right) \qquad EXP(x) = I_o e^{\frac{x}{2V_T}} - I_o, \qquad (4.49)$$

we obtain

$$EXP(\hat{V}_1) = \int \{\frac{I_o}{2V_TC} \cdot I_{in} + \frac{I_{A11}}{2V_TC} \cdot EXP(\hat{V}_1)$$

$$+ \frac{I_{A12}}{2V_TC} \cdot EXP(\hat{V}_2) + ... + \frac{I_{A1N}}{2V_TC} \cdot EXP(\hat{V}_N) \} dt \qquad (4.50)$$

Comparing this expression with the one derived from the SFG in Figure 4-4(b), i.e.,

$$EXP(\hat{V}_1) = \int [b_1 I_{in} + A_{11} EXP(\hat{V}_1) + A_{12} EXP(\hat{V}_2) + \ldots + A_{1N} EXP(\hat{V}_N)] dt$$

(4.51)

we obtain the following relationships

$$I_o = b_1 \cdot 2V_T C$$

(4.52)

and

$$I_{A_{11}} = A_{11} \cdot 2V_T C$$

$$I_{A_{12}} = A_{12} \cdot 2V_T C$$

$$\ldots\ldots\ldots$$

$$I_{A_{1N}} = A_{1N} \cdot 2V_T C$$

(4.53)

Therefore, the bias currents of the input cell and the log-domain integrators are selected according to (4.52) and (4.53) to effectively implement all coefficients in the first row of the state-space formulation.

In a similar fashion, the above procedures (i.e., the derivations from (4.44) to (4.53)), can be applied to the rest of the row equations in (4.43). Consequently, the log-domain implementation of state-space equation (4.43) will result in the circuit shown in Figure 4-6. This array of log-domain building blocks closely resembles its mathematical matrix structure. The following observations are made:

- There are N columns of log-domain circuits; each corresponds to a row in the state-space formulation. There exist N capacitive nodes on which N log-domain voltages ($\hat{V}_1$ to $\hat{V}_N$) are computed. These voltages are the log-domain representation of the state-variables, according to (4.46).

- The state-space matrix A is implemented by the circuits surrounded by a broken line which are composed of N^2 log-domain integrators (damped or undamped). The parameter A_{ij} is implemented by the corresponding log-domain integrator with bias current $I_{A_{ij}}$, which performs log-domain integration on voltage $\hat{V}_j$ to form $\hat{V}_i$. If we define a current matrix A_I as

$$A_I = \begin{bmatrix} I_{A_{11}} & I_{A_{12}} & \cdots & I_{A_{1N}} \\ I_{A_{21}} & \cdots & \cdots & \vdots \\ \vdots & \vdots & \vdots & \vdots \\ I_{A_{N1}} & \cdots & \cdots & I_{A_{NN}} \end{bmatrix}$$

(4.54)

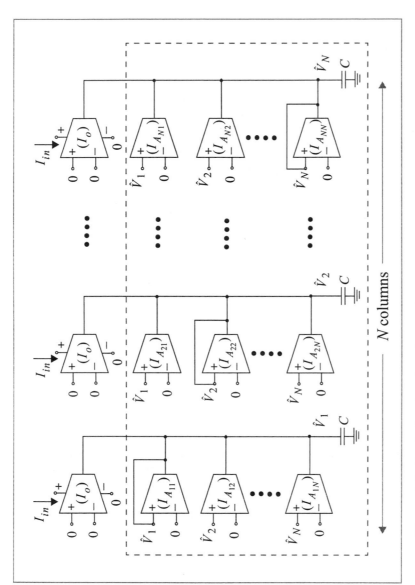

Figure 4-6: Equivalent log-domain circuit implementation of Eq. (4.43).

the log-domain bias currents are related to the state-space coefficients by

$$A_I = 2V_TC \cdot A \qquad (4.55)$$

which is based on the results of (4.53).

•   Damped log-domain integrators are recognized as the blocks having equal-indexed bias currents, such as $I_{A_{11}}, I_{A_{22}}, ..., I_{A_{NN}}$. These N cells are located along the diagonal of the shaded region.

•   The input and integration functions (as governed by the state-space vector b) are realized by the first row of circuits from the top of Figure 4-6. They perform the input LOG operation and log-domain integration as described in Table 4-2. However, they all implement the same multiplicative factor. From (4.52), the vector b is related to the current by

$$\begin{bmatrix} b_1 \\ b_2 \\ \vdots \\ b_N \end{bmatrix} = \frac{1}{2V_TC} \cdot \begin{bmatrix} I_o \\ I_o \\ \vdots \\ I_o \end{bmatrix} \qquad (4.56)$$

Consequently, the b coefficients are not individually controllable by bias currents, and they have to be set equal (or zero). This reduces the generality of the filters that can be implemented. However, the realization is still capable of implementing arbitrary filter transfer functions. This stems from the fact that many different combinations of state-space coefficients will yield the same transfer function. Later examples will demonstrate this claim. If this approach is not acceptable, then one can resort to the input stage shown in Figure 2-5(b) where an additional stage is used. This additional stage will provide the variable gain control.

•   Lastly, notice that the circuit on Figure 4-6 is only a simple illustration for a state-space system with all-positive coefficients. If matrices A or b contain negative elements, then a negative integrator should be used instead of a positive one. This is illustrated in Figure 4-7, in which the log-domain circuit realizing the coefficient A_{ij} is shown. If A_{ij} is positive, the circuit of Figure 4-7(a) is employed as usual, which is simply a positive log-domain integrator. On the other hand, a negative A_{ij} can be realized by a negative log-domain integrator (with the same bias current) as shown in Figure 4-7(b). This argument is also applicable to other log-domain building blocks. We should point out here that most, if not all, IC implementations will be constructed from fully-differential cells in order to minimize the effect of substrate noise. As a result, one can interchange the connections at the output as well as the input.

To complete the log-domain state-space filter, the summation equation (i.e., the second equation) of (4.6) needs to be realized using log-domain circuits. It is recaptured and written in vector form as shown below,

$$y = c^T x + du = \begin{bmatrix} c_1 & c_2 & \dots & c_N \end{bmatrix} \begin{bmatrix} x_1 \\ x_2 \\ \vdots \\ x_N \end{bmatrix} + du \qquad (4.57)$$

The corresponding log-domain circuit should restore the overall system linearity by performing the *EXP* operations, and realizing a weighted summation of the resulting $N+1$ linear currents. The SFG representation of this expression is provided in Figure 4-8(a) and the corresponding log-domain equivalent SFG is shown in Figure 4-8(b). We also add the following equivalence to the variable set established earlier,

$$y \Leftrightarrow I_{out}$$

Finally, the log-domain circuit implementation is shown in Figure 4-8(c). From the expressions provided in Table 4-2, the output current I_{out} is written by inspection to be

$$I_{out} = I_{in} + \left[I_{c_1} e^{\frac{\hat{V}_1}{2V_T}} - I_{c_1} \right] + \left[I_{c_2} e^{\frac{\hat{V}_2}{2V_T}} - I_{c_2} \right] + \dots + \left[I_{c_N} e^{\frac{\hat{V}_N}{2V_T}} - I_{c_N} \right] \qquad (4.58)$$

It can be equivalently written as

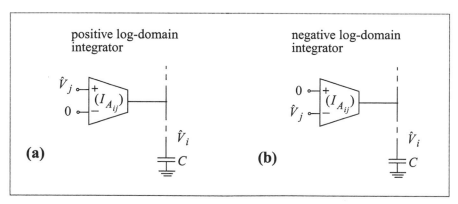

(a) positive log-domain integrator   **(b)** negative log-domain integrator

Figure 4-7: Realization of (a) a positive and (b) a negative state-space coefficient A_{ij}.

$$I_{out} = I_{in} + \frac{I_{c_1}}{I_o} \cdot \left[I_o e^{\frac{\hat{V}_1}{2V_T}} - I_o \right] + \frac{I_{c_2}}{I_o} \cdot \left[I_o e^{\frac{\hat{V}_2}{2V_T}} - I_o \right] + \dots + \frac{I_{c_N}}{I_o} \cdot \left[I_o e^{\frac{\hat{V}_N}{2V_T}} - I_o \right]$$

(4.59)

Using the transformation of (4.49), it can be simplified to be

$$I_{out} = I_{in} + \frac{I_{c_1}}{I_o} \cdot EXP(\hat{V}_1) + \frac{I_{c_2}}{I_o} \cdot EXP(\hat{V}_2) + \dots + \frac{I_{c_N}}{I_o} \cdot EXP(\hat{V}_N) \qquad (4.60)$$

Comparing this expression with the one derived from the SFG in Figure 4-8(b), i.e.,

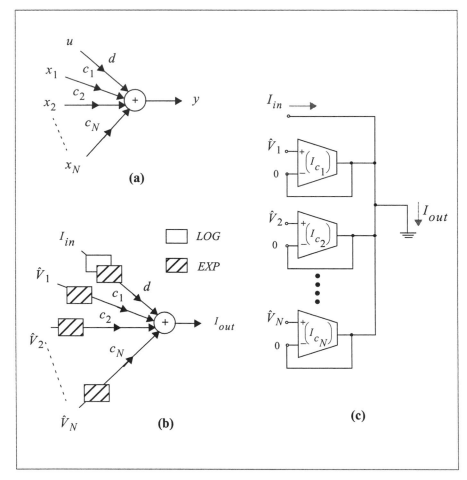

Figure 4-8: SFGs and log-domain equivalent circuit: (a) output stage SFG, (b) equivalent log-domain SFG, and (c) log-domain circuit.

$$I_{out} = d \cdot I_{in} + c_1 \cdot EXP(\hat{V}_1) + c_2 \cdot EXP(\hat{V}_2) + \ldots + c_N \cdot EXP(\hat{V}_N) \qquad (4.61)$$

and defining a current vector c_I as

$$c_I = \begin{bmatrix} I_{c_1} \\ I_{c_2} \\ \vdots \\ I_{c_N} \end{bmatrix} \qquad (4.62)$$

we find

$$c_I = I_o \cdot c \qquad (4.63)$$

and $d = 1$. In other words, the state-space parameters described by vector c will be directly controlled by the N bias currents I_{c_1} to I_{c_N}. The d element is not scalable as unity is always assigned. However, d can be set to zero by removing the I_{in} branch on Figure 4-8(c). In fact, it is always preferable to have $d = 0$ as filters with at least one transmission zero at infinity are more likely to be implemented.

Combining the integration (Figure 4-6) and the summation (Figure 4-8(c)) sections, the complete N^{th}-order state-space log-domain circuit of Figure 4-9 results. The mesh-like log-domain voltage connectivity is explicitly shown. The four regions as defined by the broken lines represent the four circuit groups that realize the state-space parameters A, b, c and d. Collectively, it performs the following log-domain signal processing operations:

$$\frac{d}{dt} \begin{bmatrix} EXP(\hat{V}_1) \\ EXP(\hat{V}_2) \\ \vdots \\ EXP(\hat{V}_N) \end{bmatrix} = \frac{1}{2V_T C} \cdot A_I \cdot \begin{bmatrix} EXP(\hat{V}_1) \\ EXP(\hat{V}_2) \\ \vdots \\ EXP(\hat{V}_N) \end{bmatrix} + \frac{I_o}{2V_T C} \cdot \begin{bmatrix} I_{in} \\ I_{in} \\ \vdots \\ I_{in} \end{bmatrix}$$

$$(4.64)$$

$$I_{out} = \frac{1}{I_o} \cdot c_I^T \cdot \begin{bmatrix} EXP(\hat{V}_1) \\ EXP(\hat{V}_2) \\ \vdots \\ EXP(\hat{V}_N) \end{bmatrix} + I_{in}$$

They correspond to the linear state-space formulation of (4.6) in a one-to-one manner

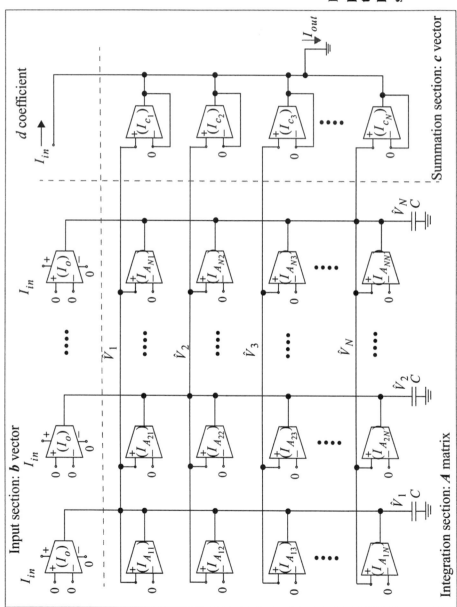

Figure 4-9: Complete circuit of the Nth-Order log-domain state-space filter.

according to equations (4.55), (4.56) and (4.63).

In summary, the design of log-domain state-space filter involves:

1. Find a good set of frequency normalized coefficients of the state-space system that meets the desired filter specification (such as the one discussed in Section 4.2.2).

2. Frequency denormalize the transfer function according to

$$A \rightarrow 2\pi f_p A \qquad b \rightarrow 2\pi f_p b \tag{4.65}$$

where f_p is the upper passband edge of the desired filter response in Hz.

3. Select a reasonable size capacitor C and bias current I_o such that

$$I_o = (2V_T C) \cdot b \tag{4.66}$$

where b is the non-zero parameters (assuming all equal) in vector b.

4. Compute the current matrices A_I and c_I using (4.55) and (4.63).

5. Employ these bias currents to realize the circuit shown in Figure 4-9.

Next, we will demonstrate the synthesis of several log-domain state-space filters to verify the practicality of the above scheme. They can all be interpreted as special cases to the Nth-order example presented previously.

4.3.2 3rd-Order Elliptic Lowpass Filter

We will implement the log-domain circuit for the low-noise low-sensitivity state-space system discussed in Section 4.2.2. The state-space coefficients are derived from a doubly terminated elliptic lowpass LC ladder. For ease of reference, the dynamic-range scaled coefficients of (4.40)-(4.41) are recaptured below:

$$A = \begin{bmatrix} -0.5212 & -0.8321 & -0.0602 \\ 0.7049 & 0 & -0.5302 \\ -0.1064 & 1.1063 & -0.5212 \end{bmatrix} \qquad b = \begin{bmatrix} 0.7840 \\ 0 \\ 0.1601 \end{bmatrix}$$

$$c = \begin{bmatrix} 0 \\ 0 \\ 1 \end{bmatrix} \qquad d = 0 \tag{4.67}$$

Due to the limitation of our log-domain implementations, it is required that the non-zero parameters of the b vector should be equal. Unfortunately, this is not the case in (4.67). To overcome this problem, the transposed state-space system as discussed in (4.9) to (4.11) can be employed. Here, we will then transpose the unscaled state equations of (4.31) and (4.32)†. From (4.11), an equivalent state-space system is written:

$$A = \begin{bmatrix} -0.5212 & 1.3297 & -0.0800 \\ -0.4411 & 0 & 0.4411 \\ -0.0800 & -1.3297 & -0.5212 \end{bmatrix} \qquad b = \begin{bmatrix} 0 \\ 0 \\ 2 \end{bmatrix}$$

$$c = \begin{bmatrix} 0.5212 \\ 0 \\ 0.0800 \end{bmatrix} \qquad d = 0$$

(4.68)

which will realize the identical filter transfer function while preserving all desirable noise and sensitivity properties. Notice that the b vector is now composed of a single non-zero element only. This is well suitable to the log-domain implementation as discussed before.

We will then scale (4.68) for maximum dynamic range. The spectral peaks for the state-variables modeled in (4.68) are computed to be 2.1404, 1.8681 and 2.7842. Following the procedures described in (4.37) and (4.38), the state-space coefficients become

$$A = \begin{bmatrix} -0.5212 & 1.1605 & -0.1041 \\ -0.5054 & 0 & 0.6575 \\ -0.0615 & -0.8922 & -0.5212 \end{bmatrix} \qquad b = \begin{bmatrix} 0 \\ 0 \\ 0.7183 \end{bmatrix}$$

$$c = \begin{bmatrix} 1.1155 \\ 0 \\ 0.2228 \end{bmatrix} \qquad d = 0$$

(4.69)

Having an implementable and optimally scaled log-domain state-space system, the next step would be to select a capacitor size. The decision usually involves a compromise between a host of conflicting interests such as silicon area, capacitor matching, linearity, noise and power. In this particular example, $C = 10$ pF is picked. For a cutoff frequency (f_p) of 100 MHz, the bias current I_o is found to be

$$\begin{aligned} I_o &= (2V_T C) \cdot b \cdot 2\pi f_p \\ &= (2 \cdot 25 \text{ mV} \cdot 10 \text{ pF}) \cdot 0.7183 \cdot 2\pi \cdot 100 \text{ MHz} \\ &= 225.67 \text{ } \mu\text{A} \end{aligned}$$

(4.70)

$\dagger$. Since a transpose manipulation will damage the equalized spectral peaks enjoyed by (4.67), an additional step of dynamic range scaling is required. In this case, we decide to derive the transpose system from the unscaled state-space equations of (4.31) and (4.32).

which is based on (4.65) and (4.66). This equation illustrates the intimate relationship between the bias current I_o and the integrating capacitor C. It is important to ensure the choice of capacitor should always yield certain convenient and reasonable bias current level.

Applying (4.55) to the frequency denormalized A matrix of (4.69), the current matrix A_I is found to be

$$A_I = \begin{bmatrix} -163.73 & 364.59 & -32.71 \\ -158.78 & 0 & 206.55 \\ -19.33 & -280.28 & -163.73 \end{bmatrix} \quad \text{in } \mu A \qquad (4.71)$$

As explained before, the current terms with negative polarity correspond to an implementation with a negative log-domain integrator. Similarly, applying (4.63) to the c vector of (4.69), the current vector c_I becomes

$$c_I = \begin{bmatrix} 251.73 \\ 0 \\ 50.29 \end{bmatrix} \quad \text{in } \mu A \qquad (4.72)$$

Therefore, the final log-domain state-space filter will appear as Figure 4-10. Notice that the zero elements of the state-space parameters are removed from the network of circuits. It is crucial to notice how the polarities of the coefficients are realized by the (voltage or current) input connections at each log-domain block. The filter transfer function will be given by the ratio of currents

$$H(s) = \frac{I_{out}}{I_{in}}(s) \qquad (4.73)$$

For a class A circuit, the input signal current I_{in} should be always smaller than the minimum bias current depicted in (4.71) and (4.72), which, in this case, equals $19\ \mu A$. Figure 4-11 shows the SPICE simulations with ideal transistors. A near-perfect 3rd-order elliptic lowpass response is achieved. Although the synthesis method is exact, slight deviation on the filter shape is still observed. This is due to the limited precision of the realizable bias currents in circuit. The current inaccuracy is directly translated to coefficient errors in the resulting state-space system. This practical limitation again stresses the importance of having a low-sensitivity state-space formulation, such as the one employed in this example.

4.3.3 3rd-Order Elliptic Highpass Filter

To show the versatility of the state-space synthesis method, a highpass filter example will be demonstrated. A 3rd-order elliptic highpass filter is designed,

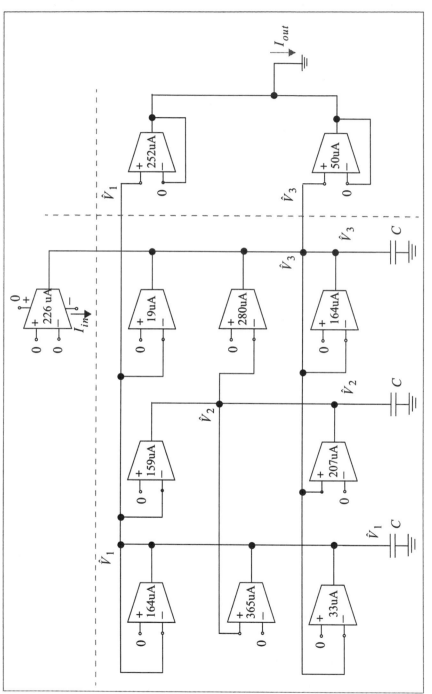

Figure 4-10: A 3rd-order log-domain lowpass elliptic filter.

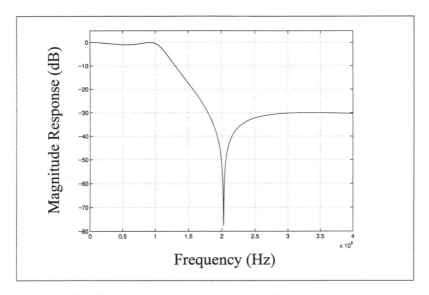

Figure 4-11: Simulated transfer function of the 3rd-order log-domain elliptic lowpass filter.

whose characteristics are described below:

$$\text{Cutoff Frequency} = 100 \text{ MHz}$$
$$\text{Passband Attenuation} = 4 \text{ dB}$$
$$\text{Stopband attenuation} = 35 \text{ dB}$$

For simplicity, a built-in filter design routine in Matlab is employed to compute the normalized state-space coefficients (with cutoff frequency of 1 rad/s),

$$A = \begin{bmatrix} -3.76 & -2.04 & -4.12 \\ 1.00 & 0 & 0 \\ 0 & 1.00 & 0 \end{bmatrix} \quad b = \begin{bmatrix} 0.16 \\ 0 \\ 0 \end{bmatrix} \quad c = \begin{bmatrix} -23.62 \\ -10.81 \\ -25.87 \end{bmatrix} \quad d = 1 \quad (4.74)$$

As b vector contains a single non-zero element, this design can be readily implemented using log-domain circuits. Assuming the capacitor size of $C = 20 \text{ pF}$, I_o is found to be 100 μA according to (4.65) and (4.66). Applying (4.55) to the denormalized A matrix of (4.74), the current matrix A_I is found to be

$$A_I = \begin{bmatrix} -2.36 & -1.28 & -2.59 \\ 0.63 & 0 & 0 \\ 0 & 0.63 & 0 \end{bmatrix} \quad \text{in mA} \quad (4.75)$$

Similarly, applying (4.63) to the c vector of (4.74), the current vector c_I becomes

$$c_I = \begin{bmatrix} -2.36 \\ -1.08 \\ -2.59 \end{bmatrix} \quad \text{in mA} \tag{4.76}$$

Unlike the previous lowpass example, the d element equals unity. Therefore, in the circuit implementation, there exists a feedforward path (with a gain of 1) directly from the filter input to the output. The complete log-domain realization is shown in Figure 4-12. The simulated frequency response is shown in Figure 4-13.

4.3.4 6th-Order Elliptic Bandpass Filter

We have discussed the design of two third-order state-space filters, realizing both lowpass and highpass responses. By the same token, high-order log-domain state-space filters can be derived without major increase in mathematical complexity. Here, we will demonstrate the design of a sixth-order bandpass filter with the following set of normalized state-space coefficients:

$$A = \begin{bmatrix} -0.809 & 0 & 0 & 6.252 & 0 & 0 \\ 1.257 & -0.406 & -1.277 & 0 & 6.252 & 0 \\ 0 & 1.277 & 0 & 0 & 0 & 6.252 \\ -6.252 & 0 & 0 & 0 & 0 & 0 \\ 0 & -6.252 & 0 & 0 & 0 & 0 \\ 0 & 0 & -6.252 & 0 & 0 & 0 \end{bmatrix} \quad b = \begin{bmatrix} 1 \\ 0 \\ 0 \\ 0 \\ 0 \\ 0 \end{bmatrix} \tag{4.77}$$

and

$$c = \begin{bmatrix} 0.403 \\ -0.130 \\ 0.413 \\ 0 \\ 0 \\ 0 \end{bmatrix} \quad d = 0 \tag{4.78}$$

Similar to the previous cases, assuming a center cutoff frequency of 100 MHz and capacitors of 20 pF, the current matrices (in μA) are routinely found to be:

$$A_I = \begin{bmatrix} -80.9 & 0 & 0 & 625.2 & 0 & 0 \\ 125.7 & -40.6 & -127.7 & 0 & 625.2 & 0 \\ 0 & 127.7 & 0 & 0 & 0 & 625.2 \\ -625.2 & 0 & 0 & 0 & 0 & 0 \\ 0 & -625.2 & 0 & 0 & 0 & 0 \\ 0 & 0 & -625.2 & 0 & 0 & 0 \end{bmatrix}; \quad c_I = \begin{bmatrix} 40.3 \\ -13.0 \\ 41.3 \\ 0 \\ 0 \\ 0 \end{bmatrix} \tag{4.79}$$

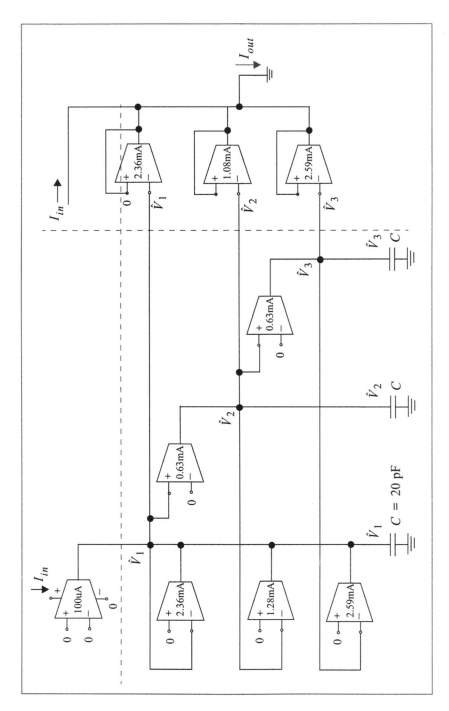

Figure 4-12: A 3rd-order log-domain highpass elliptic filter.

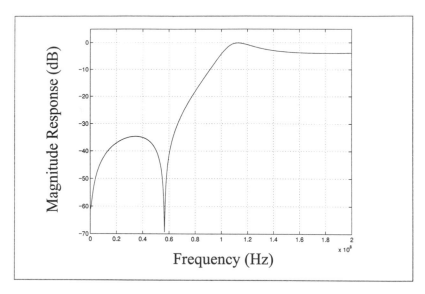

Figure 4-13: Simulated transfer function of the 3rd-order log-domain elliptic highpass filter.

Finally, the log-domain circuit, as well as the simulated filter response, are shown in Figure 4-14 and Figure 4-15, respectively. Notice that the steps involved in this sixth-order example are identical to those used in the lower-order ones, a direct result of the systematic nature of our synthesis procedure.

The method of state-space synthesis does not come without a problem. One well-known drawback is the wide range of state-space coefficient values. In conventional linear filtering technologies such as active-RC, this will translate to a wide spread of resistor and capacitor sizes. In the log-domain, this translates into a wide range of bias currents. Our readers may have already noticed this limitation in several of our previous examples. For instance, bias currents whose magnitudes differ by a decade are simultaneously required to implement the sixth-order bandpass function. According to (4.79), the smallest current is computed to be 13 μA, while the biggest one reaches 625 μA. A possible solution is provided in [32]-[33], where dedicated current sources programmable by 8-bit digital-to-analog converters (DACs) are used.

4.4 Summary

The state-space synthesis method is a general but straightforward technique for filter realization. In order to obtain an optimal set of state-space coefficients, which will lead to a low-noise and low-sensitivity design, a procedure is given based on the circuit equations of an *LC* ladder prototype. Using the log-domain

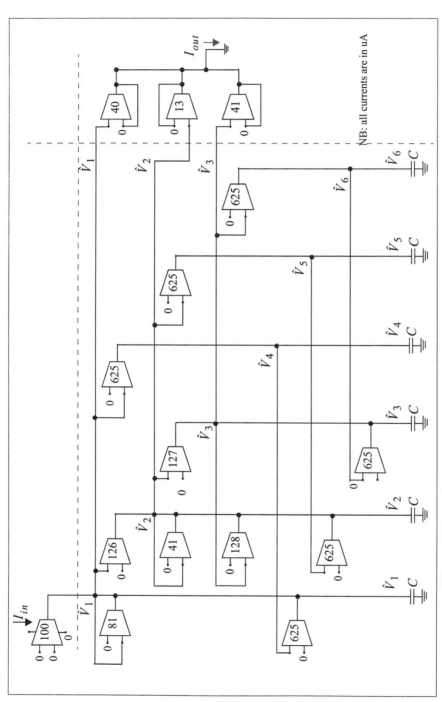

Figure 4-14: A 6th-order log-domain bandpass elliptic filter.

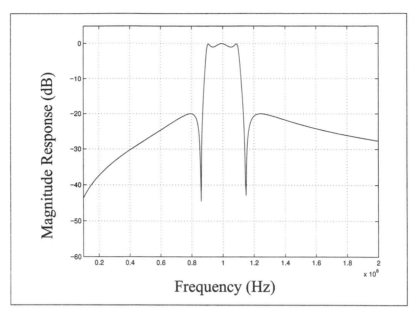

Figure 4-15: Simulated transfer function of the 6th-order log-domain elliptic bandpass filter.

integrators as building blocks, a systematic approach for realizing an arbitrary N^{th}-order state-space log-domain filter is shown. In addition, circuit examples are given which yield third- or sixth-order lowpass, highpass or bandpass elliptic filter responses.

Up to this point, we have assumed ideal bipolar devices. As the synthesis procedure is exact, ideal filter responses result. In reality, bipolar transistors suffer from an array of nonidealities, such as finite beta, non-zero ohmic junction resistances, Early voltages, component mismatches, etc. Each of these factors prevent the actual devices from exhibiting near-perfect exponential characteristics, thus filter response deviations are to be expected. We will therefore devote the next two chapters to investigate these nonideal effects in greater depth.

CHAPTER 5   Nonideality Analysis of Biquadratic Log-Domain Filters

Log-domain filters suffer directly from transistor-level nonidealities. We believe that before a new filtering technique can be industrially applied, a thorough and systematic understanding of its nonideal behavior is indispensable [58]-[59]. It enables a designer to predict the deviations in advance, so that he/she can over-design the filter to allow margins for performance variations. Moreover, nonideality analysis is the pre-requisite of filter compensation and on-chip automatic tuning.

As demonstrated in Chapter 3, loss of filter Q (or passband ripple) and shift of the cutoff (or center) frequency are direct results of transistor imperfections. The study of these nonideal effects is complicated by the non-linear logarithmic-exponential operations inherent in the circuit, resulting in complicated transcendental equations. Simplifying (but valid) assumptions are inevitably required to obtain closed-form solutions. This chapter develops simple formulae that describe the nonideal frequency operation of log-domain filters. These formulae offer practical insights into the underlying cause of these frequency distortion mechanisms and, in short, answer the question: "*Who* is causing *What*, and by *How Much*?"

The log-domain lowpass biquad shown in Figure 3-7 is chosen as our running example in this chapter, primarily due to its simplicity and sufficiency to reveal the underlying principles. Here, we follow a bottom-up approach. To begin with, the log-domain cells will be analyzed under a particular device imperfection. We will then see how the nonideality affects the operation of the log-domain integrator, and consequently, the log-domain biquad. Analytical equations will be derived that account for this deviation. As a result of this analysis, simple methods to compensate for the effects of transistor nonidealities will be proposed. The results are extended to include a bandpass biquadratic filter. In addition, some comments on the effects of transistor parasitics on the nonlinear behavior of log-domain filters in general will be given. To close the chapter, a discussion on the effects of transistor area mismatches will be given, together with a statistical analysis.

5.1 Effects of Transistor Parasitics & Compensation Methods

This section will investigate the effects of transistor parasitics such as emitter and base resistance, finite base current and Early Voltage. This section will also describe ways in which to minimize their effect.

5.1.1 Parasitic Emitter Resistances (RE)

Parasitic emitter resistance (RE) is often a major limitation to translinear circuit accuracy, particularly in high-frequency transistor processes using polysilicon emitters [38]. From a bipolar transistor's point of view, RE increases its effective base-emitter voltage according to

$$V_{BE} = V_T \ln\left(\frac{I_o}{I_S}\right) + R_E I_o \tag{5.1}$$

where I_o is the device current. With the extra term $R_E I_o$, (5.1) is clearly a departure from a purely translinear relationship given by (1.19). In this section, we would like to investigate its impact on the log-domain biquadratic filter response.

We will begin our investigation with the log-domain cells (or the so-called voltage-programmable current mirror). In the presence of RE as shown in Figure 5-1, a KVL equation can be routinely derived around the translinear loop to be

$$\hat{V}_i + (R_{EN} + R_{EP})(I_o - I_{out}) - 2V_T \ln\left(\frac{I_{out}}{I_o}\right) - \hat{V}_o = 0 \tag{5.2}$$

where R_{EN}, R_{EP} are the emitter resistances of the npn and pnp transistors, respectively. Since (5.2) is a mixture of linear and logarithmic terms, there is no explicit solution for I_{out}. However, a good approximation[†] can be obtained by performing a Taylor Series expansion, and dropping the higher-order terms, as follows:

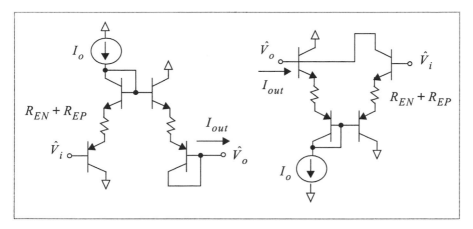

Figure 5-1: Log-domain cells showing the parasitic emitter resistances.

$$\exp(x) = 1 + x + \frac{x^2}{2!} + \frac{x^3}{3!} + \dots \approx 1 + x$$

$$\ln(x) = (x-1) - \frac{(x-1)^2}{2} + \frac{(x-1)^3}{3} - \dots \approx x - 1$$

(5.3)

We can then approximate (5.2) by

$$\hat{V}_i + (R_{EN} + R_{EP})(I_o - I_{out}) - 2V_T\left(\frac{I_{out}}{I_o} - 1\right) - \hat{V}_o = 0$$

(5.4)

Rearranging, we have

$$\left[\frac{2V_T + (R_{EN} + R_{EP})I_o}{I_o}\right]I_{out} = 2V_T + (R_{EN} + R_{EP})I_o + \hat{V}_i - \hat{V}_o$$

(5.5)

or

---

†. The validity of this approximation can be evaluated by the left-over term ξ (i.e. the error incurred) when (5.7) is substituted back into (5.2), which equals

$$\xi = (R_{EN} + R_{EP})I_o \cdot \sum_{n=2}^{\infty} \frac{1}{n!}\left(\frac{\hat{V}_i - \hat{V}_o}{2V_T + (R_{EN} + R_{EP})I_o}\right)^n$$

For 80% modulation index and $R_{EN} = R_{EP} = 20\Omega$, SPICE transient simulation reveals that the worst-case ξ converges to 1.05mV, or only 4% of V_T, thus confirming the practicality of (5.7). Its accuracy will be further confirmed when the theoretical filter deviation predicted from (5.7) is compared directly to simulation results.

$$I_{out} = I_o \left[1 + \frac{\hat{V}_i - \hat{V}_o}{2V_T + (R_{EN} + R_{EP})I_o} \right] \tag{5.6}$$

so that a convenient expression can then be obtained for I_{out} (after reversing the approximation given in (5.3)) as follows

$$I_{out} \approx I_o e^{\frac{\hat{V}_i - \hat{V}_o}{2V_T + (R_{EN} + R_{EP})I_o}} \tag{5.7}$$

Based on (5.7), the nonideal function of the log-domain integrator in Figure 2-2, which is composed of a capacitor and the log-domain cells just discussed, is now re-derived (for simpler notation, we define $2R_E = R_{EN} + R_{EP}$) by writing KCL at the capacitor node as follows

$$C \cdot \frac{d\hat{V}_o}{dt} = I_o e^{\frac{\hat{V}_{ip} - \hat{V}_o}{2V_T + 2R_E I_o}} - I_o e^{\frac{\hat{V}_{in} - \hat{V}_o}{2V_T + 2R_E I_o}} \tag{5.8}$$

Multiplying through by $e^{\hat{V}_o/(2V_T + 2R_E I_o)}$, and applying the chain rule, (5.8) becomes

$$C \cdot \left(\frac{2V_T + 2R_E I_o}{I_o} \right) \cdot \frac{d}{dt} \left(I_o e^{\frac{\hat{V}_o}{2V_T + 2R_E I_o}} - I_o \right) =$$

$$\left(I_o e^{\frac{\hat{V}_{ip}}{2V_T + 2R_E I_o}} - I_o \right) - \left(I_o e^{\frac{\hat{V}_{in}}{2V_T + 2R_E I_o}} - I_o \right) \tag{5.9}$$

If we define a new set of complementary *LOG* and *EXP* mappings as[†]

$$LOG(x) = (2V_T + 2R_E I_o) \ln \left(\frac{I_o + x}{I_o} \right) \tag{5.10}$$

and

$$EXP(x) = I_o e^{\frac{x}{2V_T + 2R_E I_o}} - I_o \tag{5.11}$$

---

†. The *LOG* and *EXP* mappings here are slightly different than that in (2.4). In fact, the log-domain synthesis theory in this book will hold for any mapping operators, $f(x)$ and $g(x)$, as long as they are complementary functions, i.e., $f(g(x)) = x$.

the RE-corrupted log-domain integration function, (5.9), can be written as

$$EXP(\hat{V}_o) = \left(\frac{V_T}{V_T + R_E I_o}\right) \cdot \frac{I_o}{2V_T} \cdot \frac{1}{C} \cdot \int \{EXP(\hat{V}_{ip}) - EXP(\hat{V}_{in})\} dt \qquad (5.12)$$

Comparing (5.12) and (2.5), it can be concluded that nonzero RE introduces a *scalar error* k_{RE} to the log-domain integrator as shown in Figure 5-2(a), where k_{RE} equals

$$k_{RE} = \frac{V_T}{V_T + R_E I_o} \qquad (5.13)$$

This log-domain integrator error will therefore affect the filter as well. To illustrate this, the SFG of the nonideal log-domain lowpass biquad is shown in Figure 5-2(b). We can now re-derive the resulting nonideal transfer function from this SFG. It is found to be

$$\frac{I_{out}}{I_{in}} = \frac{1}{1 - \frac{1}{k_{RE}^2} \cdot \left(\frac{\omega}{\omega_o}\right)^2 + \frac{j}{k_{RE}} \cdot \left(\frac{\omega}{\omega_o}\right) \frac{1}{Q}} \qquad (5.14)$$

Comparing (5.14) to the ideal case (3.22) where ω_o, Q, and K are replaced

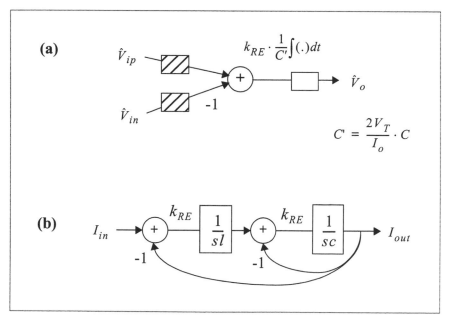

Figure 5-2: Effects of RE on: (a) log-domain integrator, and (b) the log-domain biquad.

by $\omega_{o, RE}$, Q_{RE}, and K_{RE}, respectively, the parameters of the actual filter response (cutoff frequency, filter Q and filter dc gain K) due to RE can be written as

$$\frac{\omega_{o, RE}}{\omega_o} = \frac{V_T}{V_T + R_E I_o} \qquad \frac{Q_{RE}}{Q} = 1 \qquad K_{RE} = 1 \qquad (5.15)$$

RE has the effect of lowering the actual cutoff frequency, while leaving the filter Q and gain intact. Directly from (5.15), it can be concluded that the higher the bias current, the more prominent the effect of RE. To verify this, both SPICE small-signal (AC analysis) and large-signal[†] (with 80% modulation index) simulations were performed. The results are given in Figure 5-3. As is evident, the calculated and the simulated results are in very close agreement.

The error introduced by RE can be easily absorbed by tweaking the capacitor values from C to C_{comp} which equals $k_{RE} \cdot C$. However, a more elegant electronic

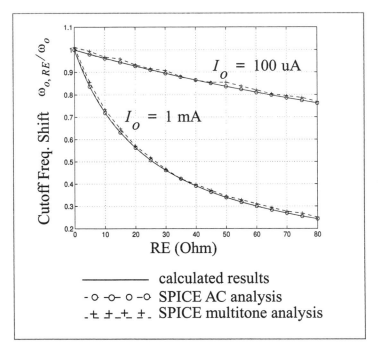

Figure 5-3: Effects of RE on filter cutoff frequency.

---

† . AC analysis is limited because it relies on linearizing the nonlinear elements of the circuit, thus neglecting its basic translinear nature. Multitone testing, which employs FFT to extract frequency responses from SPICE simulated transient signals, was used to supplement the AC results [30]. All multitone simulations presented in this chapter were performed with a modulation index of 80%.

compensation can be performed by tuning the bias current from I_o to

$$I_{comp} = \frac{V_T}{V_T - R_E I_o} \cdot I_o \qquad (5.16)$$

This is directly evident from (5.8) (see the detailed derivation in Appendix D). SPICE simulations were performed to verify this compensation scheme and are shown in Figure 5-4. Here we see two different cases for RE of 5 Ω and 20 Ω. When the current for each integrator is modified according to (5.16), the filter returns to its ideal response. For all practical purpose, I_{comp} usually represents a slight tweak of the bias current I_o dictated by the ideal synthesis procedures. For example, in a typical situation such as RE=5 Ω, and 100 μA bias current, I_o will be tuned about 2% higher to compensate for the cutoff frequency drop.

An upper limit to the amount of compensation that can be applied can be seen from the denominator of (5.16). If R_E is too large, the term $V_T - R_E I_o$ will go negative and the compensation current will change sign. Physically, this extends beyond the limits of the mathematical model and suggests that this compensation scheme will fail. For instance, given a 1-mA bias current, this scheme will fail to compensate for any R_E bigger than 25 Ω. However, even before that theoretical limit is reached, I_{comp} can grow prohibitively big as $R_E I_o$ becomes comparable to V_T. Therefore, attention should be paid to the maximum level of bias current that one is

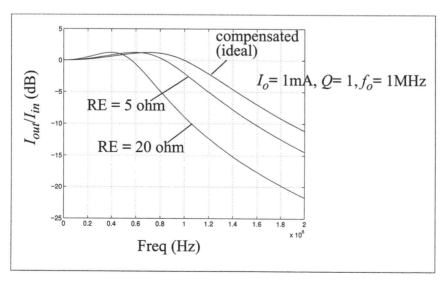

Figure 5-4: Simulated results of nonzero RE compensation.

willing to use in the design of log-domain filter with this compensation scheme.

5.1.2 Finite Beta

Following steps outlined in the previous section, we can derive the characteristic equation for the log-domain cells of Figure 2-1 subject to nonideal transistor beta effects to be as follows[†]:

$$\text{positive log-domain cell:} \quad I_{out} = \frac{I_o e^{(\hat{V}_i - \hat{V}_o)/(2V_T)}}{1 + \left(\dfrac{1}{\beta + 1}\right) \cdot e^{(\hat{V}_i - \hat{V}_o)/(2V_T)}} \tag{5.17}$$

and

$$\text{negative log-domain cell:} \quad I_{out} = \frac{I_o}{\beta + 1} + \frac{\left(\dfrac{\beta}{\beta + 1}\right) \cdot I_o e^{(\hat{V}_i - \hat{V}_o)/(2V_T)}}{1 + \left(\dfrac{1}{\beta + 1}\right) \cdot e^{(\hat{V}_i - \hat{V}_o)/(2V_T)}} \tag{5.18}$$

We can apply (5.17)-(5.18) to the log-domain integrator and write

$$
\begin{aligned}
C \cdot \frac{d\hat{V}_o}{dt} &= \frac{I_o e^{(\hat{V}_i - \hat{V}_o)/(2V_T)}}{1 + \left(\dfrac{1}{\beta + 1}\right) \cdot e^{(\hat{V}_i - \hat{V}_o)/(2V_T)}} \\
&\quad - \left(\frac{I_o}{\beta + 1} + \frac{\left(\dfrac{\beta}{\beta + 1}\right) \cdot I_o e^{(\hat{V}_i - \hat{V}_o)/(2V_T)}}{1 + \left(\dfrac{1}{\beta + 1}\right) \cdot e^{(\hat{V}_i - \hat{V}_o)/(2V_T)}} \right)
\end{aligned}
\tag{5.19}
$$

In this form, the integrator equation depicted in (5.19) is very difficult to manipulate for further analysis. Instead, we can revisit (5.17) and recognize that the log-domain cell under finite β can be viewed as a negative feedback system in which I_o and I_{out} denote the input and output, respectively. As such, the feedforward gain is $\exp[(\hat{V}_i - \hat{V}_o)/(2V_T)]$ while the feedback gain is given by $1/(\beta + 1)$ as shown in Figure 5-5(a). Notice that a similar argument also applies to (5.18). With this repre-

---

[†]. To make the analysis as simple as possible, we assume equal β for the npn and pnp transistors, which is usually not true. Nonetheless, the results presented here are still practical, because the overall performances are dominated by the most nonideal device, and we can use the results here to predict a bound on the worst-case deviation.

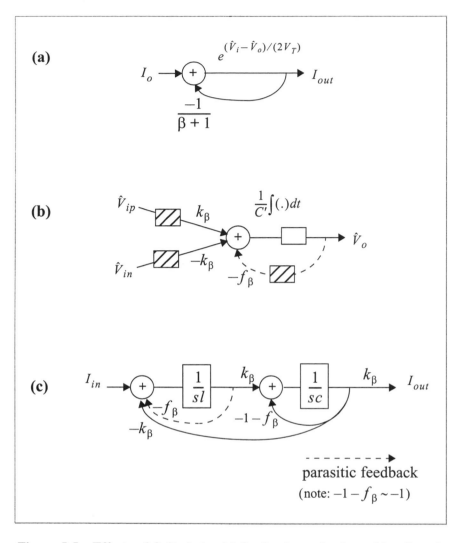

Figure 5-5: Effects of finite beta: (a) feedback mechanism of log-domain cells, (b) log-domain integrator, and (c) log-domain biquad.

sentation, we can conclude that the circuit suffers parasitic negative feedback errors that alter the function of the log-domain cell. Recognizing this, we found the following form of the equation very effective in capturing the essential behavior of the log-domain integrator,

$$C \cdot \frac{d}{dt}\hat{V}_o = k_\beta \cdot I_o e^{\frac{\hat{V}_{ip} - \hat{V}_o}{2V_T}} - k_\beta \cdot I_o e^{\frac{\hat{V}_{in} - \hat{V}_o}{2V_T}} - f_\beta \cdot I_o \qquad (5.20)$$

Here k_β accounts for any integrator scalar error and f_β denotes the negative feedback action of the log-domain cell. Based on derivations similar to those that arrived at Eqns. (2.2) to (2.5), the nonideal log-domain integration, written in terms of *LOG* and *EXP* mappings, becomes

$$EXP(\hat{V}_o) = \frac{I_o}{2V_T} \cdot \frac{1}{C} \cdot \int \{k_\beta \cdot (EXP(\hat{V}_{ip}) - EXP(\hat{V}_{in})) - f_\beta \cdot EXP(\hat{V}_o)\} dt \quad (5.21)$$

The corresponding SFG is shown in Figure 5-5(b). Having established the general form of the nonideal log-domain integrator, we can start to impose some simplifications, such as $f_\beta \ll k_\beta$ and $f_\beta \ll 1$, because, typically, $k_\beta \sim 1$ and $f_\beta \sim 0$. Incorporating the integrator errors into the biquad SFG as shown in Figure 5-5(c), the nonideal lowpass biquad transfer function is found to be

$$\frac{I_{out}}{I_{in}} = \frac{\dfrac{k_\beta^2}{k_\beta^2 + f_\beta}}{1 - \dfrac{1}{k_\beta^2 + f_\beta}\left(\dfrac{\omega}{\omega_o}\right)^2 + \dfrac{j}{k_\beta^2 + f_\beta}\left(\dfrac{\omega}{\omega_o}\right)\left(\dfrac{1}{Q} + f_\beta \cdot Q\right)} \quad (5.22)$$

Comparing (5.22) to the ideal case (3.22) where ω_o, Q, and K are replaced by $\omega_{o,\beta}$, Q_β, and K_β, respectively, the filter deviations become

$$\frac{\omega_{o,\beta}}{\omega_o} = k_\beta \qquad \frac{Q_\beta}{Q} = \frac{k_\beta}{1 + f_\beta \cdot Q^2} \qquad K_\beta = \frac{k_\beta^2}{k_\beta^2 + f_\beta} \sim 1 \quad (5.23)$$

We found the variables k_β and f_β can be approximated by

$$k_\beta = \frac{\beta}{\beta + 2} \qquad f_\beta = \left(\frac{2}{\beta + 7}\right)^2 \quad (5.24)$$

The k_β expression is very simple, which equals the conventional current mirror error. This is reasonable because the log-domain cells are essentially current mirrors [38]. The f_β expression shows that the negative feedback error is relatively small and inversely related to β. As revealed by (5.23), and confirmed by SPICE simulations, the actual cutoff frequency due to finite β will suffer a drop quantified by k_β. Although the f_β parameter is small, its effect on the filter Q is magnified by Q^2. Figure 5-6 demonstrates this by showing both the simulated and the calculated results.

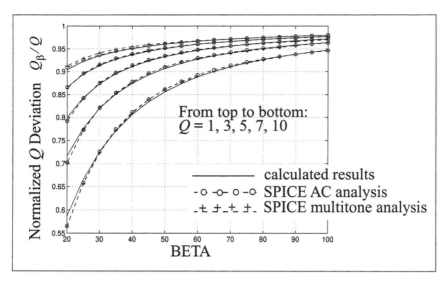

Figure 5-6: Effects of finite beta on filter Q factor.

It should be noted that the nonidealities caused by finite β can be electroni-cally compensated by: (i) tuning (increasing) the bias current from I_o to I_o/k_β to overcome the scalar errors, and (ii) pushing a dc current $f_\beta \cdot I_o$ into the non-damped integrating node to counterbalance the parasitic negative feedback. Figure 5-7(a) illustrates these two compensation techniques applied to our biquad example. SPICE simulations shown in Figure 5-7(b) verify this β compensation scheme.

For simplicity, we have assumed an invariant beta in our derivations. Gener-ally speaking, β is a function of collector current. The extent of the variation depends on the surface conditions of the fabrication. For modern process where surface recom-bination is small, β will stay substantially constant for many decades of I_C. Also, the dependency of β on temperature is relatively small and will have a negligible effect on well-designed translinear circuits [60]. It is well known that β is frequency-depen-dent as well, which can be described by a one-pole rolloff function [60],

$$\beta(s) = \frac{\beta_o}{1 + s\beta_o/\omega_T} \qquad (5.25)$$

where β_o is the beta at low frequency, and ω_T denotes the unity current gain-band-width product. In the high frequency range where log-domain filters intend to operate, we should consider this β reduction when computing biquad filter deviations using

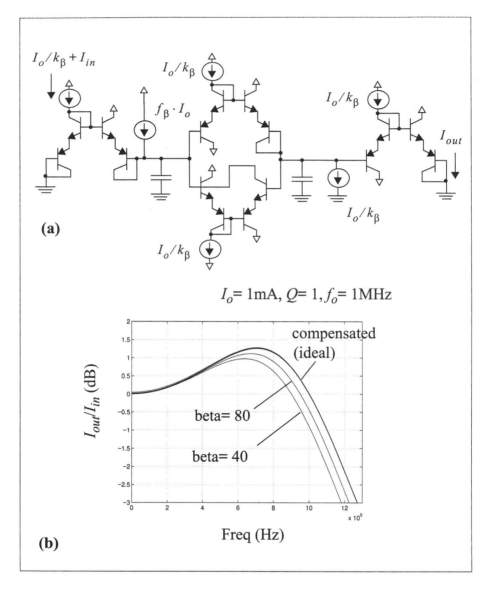

(a)

I_o= 1mA, Q= 1, f_o= 1MHz

(b)

Figure 5-7: (a) Compensation of the nonideal effects due to finite beta, and (b) the simulated results.

Eq. (5.23)-(5.24).

5.1.3 Parasitic Base Resistances (RB)

Similar to RE, RB also introduces an additional voltage drop ($R_B I_B$) across each device in the translinear loop. However, it has no effect unless there is non-zero base current, or when β is finite. Therefore, it acts as a secondary nonideal factor, acting on the beta-corrupted filter ($\omega_{o,\beta}$, Q_β) and causing more deviations. Working out its effects following Eqns. (5.2) to (5.15), we find the *scalar error* k_{RB} to be,

$$k_{RB} = \frac{V_T}{V_T + \frac{R_B I_o}{\beta}} \tag{5.26}$$

so that the biquad transfer function deviation becomes

$$\frac{\omega_{o,RB}}{\omega_{o,\beta}} = \frac{V_T}{V_T + \frac{R_B I_o}{\beta}} \qquad \frac{Q_{RB}}{Q_\beta} = 1 \qquad \frac{K_{RB}}{K_\beta} = 1 \tag{5.27}$$

where $2R_B = R_{BN} + R_{BP}$. According to (5.27), nonzero RB will further lower the cutoff frequency, whose effect is enlarged by higher bias current and lower β. Figure 5-8 provides the simulation results to support these findings.

Similar to the RE case, the effects of RB can be compensated by tuning the

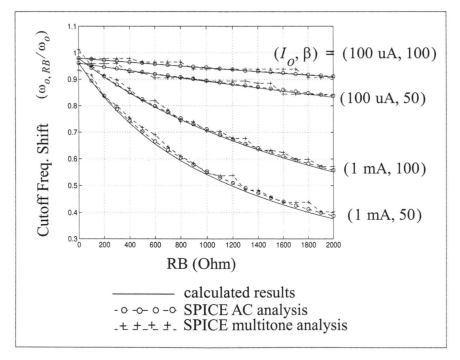

Figure 5-8: Effects of RB on filter cutoff frequency.

bias current from I_o to I_{comp} given by

$$I_{comp} = \frac{V_T}{V_T - \dfrac{R_B I_o}{\beta}} \cdot I_o \tag{5.28}$$

So far, we have assumed constant junction resistances (RE, RB). Very simple and easy-to-use deviation formulae are derived which show close agreement with SPICE simulations. However, in reality, these resistances are function of temperature and current: they will increase by about 0.15% per °C; and decrease due to emitter crowding at high currents. Therefore, as a note of caution, our assumption is only valid given small temperature fluctuation and relatively small current level. More theoretical and experimental work is necessary to fully explore (and to compensate for) this aspect of the nonideal mechanism.

5.1.4 Early Voltages

The Early effect "modulates" the effective saturation current of a bipolar device according to

$$I_{S,\,\text{eff}} = I_S \cdot \left(1 + \frac{V_{CE}}{V_A}\right) \tag{5.29}$$

For many high-frequency transistors, V_A can be quite low (say, from 5 - 50 V). In other words, collector current is heavily influenced by V_{CE}, which is a clear violation of an ideal translinear device. However, the problem may not be as severe as it may first seem. In a translinear circuit, transistors are always biased in pairs, thus mutually canceling their Early effect. The analysis presented shortly will support this argument.

In the log-domain cells of Figure 2-1, the V_{CE}'s of the transistors around the translinear loops are different; causing their effective I_S's to be mismatched. Therefore, re-deriving the log-domain equation in (2.1) results in

$$I_{out} = k_{V_A} \cdot I_o e^{\frac{\hat{V}_i - \hat{V}_o}{2 V_T}} \qquad \text{where} \qquad k_{V_A} = \sqrt{\frac{\left(1 + \dfrac{V_{CE2}}{V_{AN}}\right)\left(1 + \dfrac{V_{EC4}}{V_{AP}}\right)}{\left(1 + \dfrac{V_{CE1}}{V_{AN}}\right)\left(1 + \dfrac{V_{EC3}}{V_{AP}}\right)}} \tag{5.30}$$

This is equivalent to a bias current error or *scalar error* because, while comparing (5.30) to (2.1), I_o is now replaced by $k_{V_A} I_o$. Since the V_{CE}'s are signal dependent $(\hat{V}_i, \hat{V}_o)$, a subtle *parasitic feedback* mechanism is expected, which, as in (5.20), can be represented by the parameter f_{V_A}. Subsequently, we can investigate the Early

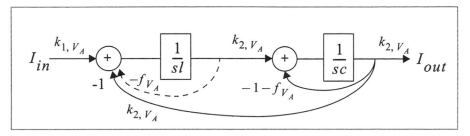

Figure 5-9: Nonideal log-domain biquad SFG due to Early effects.

effect using an analysis similar to that used to investigate the effect of finite β seen previously.

For simplicity, we assume; $V_{AN} = V_{AP} = V_A$, all V_{BE}'s (or V_{EB}'s) are equal, symmetrical power rails, and negligible log-domain voltages (at the integration nodes) relative to the power supplies[†]. A SFG incorporating the nonidealities can then be drawn as shown in Figure 5-9. Routinely re-deriving the transfer function, we have

$$\frac{I_{out}}{I_{in}}(j\omega) = \frac{\dfrac{k_{1,V_A} \cdot k_{2,V_A}^2}{k_{2,V_A}^2 + f_{V_A}}}{1 - \dfrac{1}{k_{2,V_A}^2 + f_{V_A}}\left(\dfrac{\omega}{\omega_o}\right)^2 + \dfrac{j}{k_{2,V_A}^2 + f_{V_A}}\left(\dfrac{\omega}{\omega_o}\right)\left(\dfrac{1}{Q} + f_{V_A}Q\right)} \tag{5.31}$$

Comparing (5.31) to (3.22) where ω_o, Q, and K are replaced by ω_{o,V_A}, Q_{V_A}, and K_{V_A}, we can conclude (appropriately assuming $f_{V_A} \ll k_{V_A}$)

$$\frac{\omega_{o,V_A}}{\omega_o} = k_{2,V_A} \qquad \frac{Q_{V_A}}{Q} = \frac{k_{2,V_A}}{1 + f_{V_A}Q^2} \qquad K_{V_A} = k_{1,V_A} \tag{5.32}$$

The k and f parameters can be approximated with

[†]. This is a valid assumption because, during normal filter operation, the log-domain voltages are limited to a few V_T's due to the companding nature of the circuit.

$$k_{1, V_A} = \sqrt{\dfrac{1 + \dfrac{V_{CC} - V_{BE}}{V_A}}{1 + \dfrac{V_{BE}}{V_A}}} \qquad k_{2, V_A} = \sqrt{\dfrac{1 + \dfrac{V_{CC} - V_{BE}}{V_A}}{1 + \dfrac{V_{CC} + V_{BE}}{V_A}}} \qquad (5.33)$$

$$f_{V_A} = \dfrac{15}{(V_A + 50)^2}$$

where V_{CC} is the supply voltage as shown in Figure 3-7. From the above analysis, we see that the finite Early voltage would slightly drop the actual cutoff frequency, raise the filter dc gain, and cause Q-degradation. SPICE simulations confirm the theory, as illustrated in Figure 5-10.

Finally, in much the same manner as the finite β case, compensation is achieved by simultaneously (i) tuning the bias current to correct for the cutoff frequency, and (ii) injecting a dc current into the non-damped integration node to correct for the filter Q.

Compared to the other transistor nonideal parameters discussed so far, the effects of finite V_A on the filter response are relatively mild. This is a testament to our previous argument that translinear circuits are relatively insensitive to the Early effect. The V_A's tend to mutually cancel, as the devices are arranged in pairs (see Eq. (5.30)). Nonetheless, even without compensation, the Early effect can be eliminated by careful circuit design. For example, it is possible to use a cascode stage or a separate bias line to keep the V_{CB} of each device at or near zero [38].

5.1.5 Combined Effects

The combined effects of the device model nonidealities (i.e., RE, RB, β and V_A) on the filter characteristics can be written as:

$$\Delta\omega_o \approx \frac{\partial \omega_o}{\partial R_E} \cdot R_E + \frac{\partial \omega_o}{\partial R_B} \cdot R_B + \frac{\partial \omega_o}{\partial (1/\beta)} \cdot \left(\frac{1}{\beta}\right) + \frac{\partial \omega_o}{\partial (1/V_A)} \cdot \left(\frac{1}{V_A}\right)$$

$$\Delta Q \approx \frac{\partial Q}{\partial (1/\beta)} \cdot \left(\frac{1}{\beta}\right) + \frac{\partial Q}{\partial (1/V_A)} \cdot \left(\frac{1}{V_A}\right)$$

(5.34)

where the partial derivatives are obtained by differentiating (5.15), (5.23), (5.27), (5.32) with respect to the appropriate device parameters. . The Early effect dominates the filter gain K, as given by (5.32). To verify (5.34), a multitone analysis of a biquad with a 1 MHz cutoff frequency was performed. Three different models of transistor nonidealities were employed and the results are gathered in Table 5-1. Comparing the simulated results with those predicted by hand calculations, the usefulness of (5.34) is

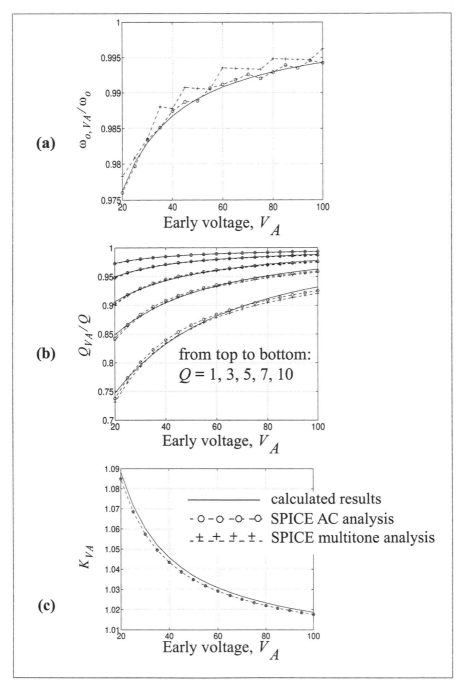

Figure 5-10: Effects of finite Early voltage on: (a) cutoff frequency, (b) filter Q; and (c) filter gain K.

Tx models	SPICE Simulations				Calculated Results			
$R_E(\Omega)$, $R_B(K\Omega)$, β, $V_A(V) =$	f_o (MHz)	$Q=1$ Q	$Q=5$ Q	K	f_o (MHz)	$Q=1$ Q	$Q=5$ Q	K
i (5, 0.15, 100, 100)	0.952	0.974	4.735	1.016	0.951	0.974	4.693	1.019
ii (20, 1, 80, 80)	0.865	0.970	4.569	1.017	0.859	0.968	4.605	1.023
iii (40, 2, 50, 50)	0.739	0.956	4.144	1.020	0.725	0.950	4.357	1.037

Table 5-1: Combined effects of device nonidealities.

indeed confirmed.

5.2   Extension to Bandpass Biquad Filters

The analysis for the lowpass biquad can be applied directly to the nonideal bandpass case. We will focus on the bandpass biquad example described in Section 3.2.2. By routine calculations, which will not be repeated here, the filter deviations are derived and summarized in Table 5-2.

Identical to the lowpass biquad, the bandpass filter nonidealities also stem from the log-domain integrator scalar and parasitic feedback errors. Very similar deviation equations are observed, thus again confirming the soundness of the previous analysis. By the same token, the filter center frequency and the Q (bandpass shape) can be independently compensated by bias current tuning and parasitic feedback cancellation.

For high frequency applications (up to 100 MHz range), new log-domain bandpass biquads have recently emerged [17]-[18]. These circuits share the same structure, except that the slow pnp transistors of the log-domain integrator are now skillfully eliminated, resulting in a high-speed all-npn realization. Nevertheless, as SPICE simulation reveals, the design equations presented in Table 5-2, with minor modifications of the variables, are still generally correct to model their deviations.

5.3   Comments on Distortion

We have explained and quantified the *linear* transfer function deviations suffered by log-domain biquads due to transistor nonidealities. Nonideal log-domain cell and integrator expressions are derived which are then used to predict the final filter deviations. At the same time, these equations also shed light on the *non-linear*, or distortion, mechanism experienced by the log-domain circuit. This brief section will qualitatively discuss several distortion observations. SPICE simulations on the lowpass biquad are also provided for illustration.

One of the most noticeable sources of distortion known to translinear circuit design is the junction resistance. We have seen that emitter resistance RE corrupts the ideal log-domain cell equation, adding an unsolvable linear term, $2R_E(I_o - I_{out})$, to the otherwise purely logarithmic equation as shown in (5.2), which is recaptured below,

$$\hat{V}_i + (2R_E)(I_o - I_{out}) - 2V_T \ln\left(\frac{I_{out}}{I_o}\right) - \hat{V}_o = 0 \qquad (5.35)$$

where $2R_E = R_{EN} + R_{EP}$. Although we have employed a Taylor series expansion to solve this equation and came up with a new *LOG/EXP* mapping, our approach is only an approximation. In fact, (5.35) dictates that under non-zero RE, the log-domain cell,

Source of Errors	Bandpass Filter Performance Deviations			Remarks
	f_{oa}/f_o	Q_a/Q	K	
RE, RB	$\dfrac{V_T}{V_T + \left(R_E + \dfrac{R_B}{\beta}\right)I_o}$	1	1	major cause of center freq. shift.
Finite β	k_β	$\dfrac{k_\beta^{3/2}}{1+f_\beta \cdot Q^2}$	$\dfrac{k_\beta^2}{1+f_\beta \cdot Q^2}$	k_β and f_β as defined in (5.24)
Early Effect (V_A)	k_{2,V_A}	$\dfrac{k_{2,V_A}}{1+f_{V_A} Q^2}$	$\dfrac{k_{1,V_A} \cdot k_{2,V_A}}{1+f_{V_A} \cdot Q^2}$	k_{1,V_A}, k_{2,V_A} and f_{V_A} as defined in (5.33)
Area Mismatches	$\sqrt[4]{\gamma_2\gamma_3}$	$\sqrt[4]{\gamma_2\gamma_3}$	$\sqrt{\gamma_1\gamma_4}$	γ's as defined in (5.38) and Figure 3-11, to be discussed shortly

Table 5-2: Summary of log-domain bandpass biquad deviations.

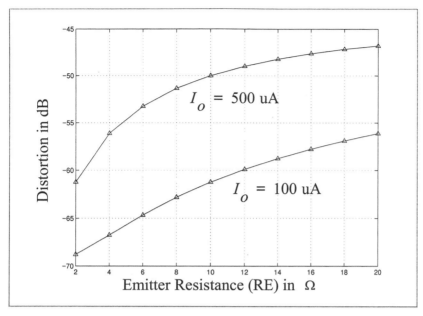

Figure 5-11: Effects of emitter resistance on log-domain biquad distortion.

and the resulting integrator, will no longer follow the exact *LOG/EXP* cancellation, thus giving rise to distortion. In fact, according to (5.35), we can assert that distortion will increase with bigger RE and bias current. To confirm this, Figure 5-11 displays the SPICE simulated[†] distortion performance of the lowpass biquad filter. It shows that the distortion rises with higher RE and I_o.

On the contrary, the effects of finite β and V_A on distortion are relatively small. As revealed by Eq. (5.17) and (5.30), although finite β and V_A introduce unwanted scaling and feedback errors to the log-domain cell, they do not, at least according to the first-order circuit analysis, corrupt the underlying *LOG/EXP* operation. SPICE simulations show a constant level of distortion as β and V_A is decreased.

However, as a secondary effect, finite β can cause significant distortion in the presence of base resistance RB. As discussed earlier, the effects of the voltage drop $R_B I_o / \beta$ are very comparable to that of $R_E I_o$. An equation similar to (5.35) can be written for RB as follows

---

† To simulate the distortion performance on SPICE, two tones are applied close to the passband edge to observe the resulting inter-modulation distortion. A modulation index of 80% was employed.

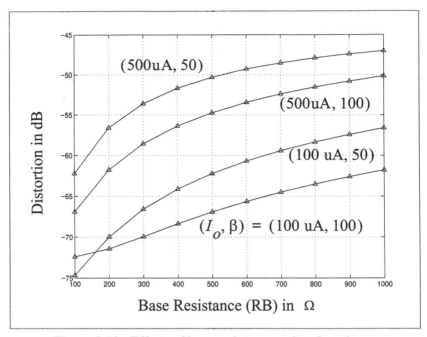

Figure 5-12: Effects of base resistance on log-domain biquad distortion.

$$\hat{V}_i + \left(\frac{2R_B}{\beta}\right)(I_o - I_{out}) - 2V_T \ln\left(\frac{I_{out}}{I_o}\right) - \hat{V}_o = 0 \qquad (5.36)$$

where $2R_B = R_{BN} + R_{BP}$. Similarly, one would expect that distortion will increase with bigger RB, higher I_o or smaller β. Figure 5-12 displays the simulated distortion versus RB at different levels of I_o and β, whose trend agrees well with our intuition. Finally, we observe that when comparing Figure 5-12 with Figure 5-11, a similar level of distortion is observed under comparable level of nonideality, i.e. $R_E I_o \sim R_B I_o / \beta$. This further confirms our claim that the two junction resistances, RE and RB, are causing distortion in the log-domain biquad through a very similar mechanism.

5.4 Transistor Area Mismatches

Ideal log-domain filtering assumes all transistors have equal areas, hence, the same saturation currents. However, this is hardly true in practice. Defining the area mismatch parameter[†], γ, to be the ratio of I_S's in the 4-transistor translinear loop (Figure 2-1), the ideal log-domain equation (2.1) will change to

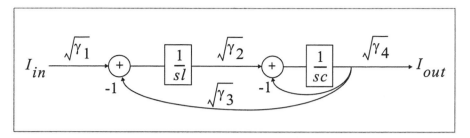

Figure 5-13: Nonideal log-domain biquad SFG due to area mismatches.

$$I_{out} = \sqrt{\gamma} \cdot I_o e^{\frac{\hat{V}_i - \hat{V}_o}{2V_T}} \qquad \text{where} \qquad \gamma = \frac{I_{S2}I_{S4}}{I_{S1}I_{S3}} \qquad (5.37)$$

Clearly, area mismatch can be equivalently viewed as bias current error, or *scalar error*. Similarly, the log-domain integrators will also suffer the same scalar error. Since there are four translinear loops in the biquad filter (Figure 3-7), we need to define four γ's to account for all possible mismatches because they are not necessarily equal, which we rewrite as

$$\gamma_i = \frac{I_{S(4i-2)}I_{S(4i)}}{I_{S(4i-3)}I_{S(4i-1)}} \qquad \text{for } i = 1 \text{ to } 4. \qquad (5.38)$$

The biquad SFG is then redrawn in Figure 5-13, from which the nonideal filter response is derived,

$$\frac{I_{out}}{I_{in}}(j\omega) = \frac{\sqrt{\frac{\gamma_1 \gamma_4}{\gamma_3}}}{1 - \frac{1}{\sqrt{\gamma_2 \gamma_3}}\left(\frac{\omega}{\omega_o}\right)^2 + \frac{j}{\sqrt{\gamma_2 \gamma_3}}\left(\frac{\omega}{\omega_o}\right)\left(\frac{1}{Q}\right)} \qquad (5.39)$$

We can then conclude

$$\frac{\omega_{o,\gamma}}{\omega_o} = \sqrt[4]{\gamma_2 \gamma_3} \qquad \frac{Q_\gamma}{Q} = \sqrt[4]{\gamma_2 \gamma_3} \qquad K_\gamma = \sqrt{\frac{\gamma_1 \gamma_4}{\gamma_3}} \qquad (5.40)$$

To verify this result, the log-domain biquad is simulated under four different mismatch conditions as shown in Table 5-3. Cases (i)-(iii) represent random mis-

---

†. In the literature, area mismatch is also known as: "saturation current mismatch", "V_{BE} mismatch", or "equivalent offset voltage".

Mis-matches	Small-signal Simulation			Large-signal Simulation			Calculated Results		
	f_{oa} (MHz)	Q_a	K	f_{oa} (MHz)	Q_a	K	f_{oa} (MHz)	Q_a	K
None	0.9998	0.9991	1.0018	1.0112	0.9988	1.0018	1	1	1
case (i)	1.0045	1.0043	0.8809	1.0060	1.0039	0.8808	1.005	1.005	0.879
case (ii)	1.0719	1.0716	0.7234	1.0807	1.0712	0.7234	1.073	1.073	0.722
case (iii)	1.8192	1.8175	0.2906	1.8291	1.8147	0.2906	1.820	1.820	0.290
case (iv)	0.9998	0.9991	1.0016	1.0112	0.9988	1.0016	1.000	1.000	1.000

Table 5-3: Filter performances under different mismatch conditions.

$(\gamma_1, \gamma_2, \gamma_3, \gamma_4) =$

Case (i) : (0.856, 0.924, 1.105, 0.998);
Case (ii) (0.524, 0.683, 1.938, 1.928);
Case (iii) (1.992, 3.738, 2.938, 0.124);
Case (iv) (5.908, 0.237, 4.219, 0.714).

matches with increasing severity, while case (iv) shows a mutually cancelled mismatch situation. Good agreement between the SPICE simulations and the values predicted from (5.40) are observed. Interestingly, case (iv) demonstrates that virtually no filter deviations will occur given the mismatches exactly cancel each other.

5.4.1 Statistical Analysis Of Area Mismatches

The previous comparisons can only serve as a quick verification of the analysis. Very seldom can we know the mismatches pre-deterministically. Since transistor mismatches are results of variations in the fabrication process, they are random in nature, and are best studied using statistical methods.

By the *transmission of moment's formula* [61], if Y is a function of N independent, Gaussian random variables (r.v.) X_i (with mean μ_{X_i} and variance $\sigma_{X_i}^2$), then the mean and variance of Y can be approximated by

$$\mu_Y = f(\mu_{X_1}, \mu_{X_2}, \ldots\ldots, \mu_{X_N}) \qquad \sigma_Y^2 = \sum_{i=1}^{N} \left(\frac{\partial f}{\partial x_i}\right)^2 \sigma_{X_i}^2 \qquad (5.41)$$

where the partial derivatives are evaluated at μ_{X_i}. The smaller the $\sigma_{X_i}^2$, the more accurate the expression becomes. We will apply (5.41) to analytically derive the statistical filter response deviations caused by area mismatches.

To begin with, we assume that the I_S of the ith npn and pnp transistors are governed by:

$$I_{SN,i} = I_{SN,\,nominal}(1 + X_{N,i}) \qquad I_{SP,i} = I_{SP,\,nominal}(1 + X_{P,i}) \qquad (5.42)$$

respectively, where $X_{N,i}$ and $X_{P,i}$ are *independent* and *Gaussian* r.v.'s, each having zero mean and $\sigma_{X_N}^2$ (or $\sigma_{X_P}^2$) variance. Rewriting (5.40) in terms of the corresponding I_S and r.v.'s, we have

$$\frac{\omega_{o,Y}}{\omega_o} = \frac{Q_Y}{Q} = \sqrt[4]{\frac{I_{S6}I_{S8}I_{S10}I_{S12}}{I_{S5}I_{S7}I_{S9}I_{S11}}}$$

$$= \sqrt[4]{\frac{(1 + X_{N6})(1 + X_{P8})(1 + X_{N10})(1 + X_{P12})}{(1 + X_{N5})(1 + X_{P7})(1 + X_{N9})(1 + X_{P11})}} \qquad (5.43)$$

$$K_\gamma = \sqrt{\frac{I_{S2}I_{S4}I_{S9}I_{S11}I_{S14}I_{S16}}{I_{S1}I_{S3}I_{S10}I_{S12}I_{S13}I_{S15}}}$$

$$= \sqrt{\frac{(1+X_{N2})(1+X_{P4})(1+X_{N9})(1+X_{P11})(1+X_{N14})(1+X_{P16})}{(1+X_{N1})(1+X_{P3})(1+X_{N10})(1+X_{P12})(1+X_{N13})(1+X_{P15})}}$$

(5.44)

Applying the transmission of moment's formula (5.41) to Eqns. (5.43) to (5.44), the variances of the filter deviations are as follows,

$$\sigma^2_{\omega_{o,\gamma}/\omega_o} = \sigma^2_{Q_{o,\gamma}/Q} = \frac{1}{4}\cdot(\sigma^2_{X_N}+\sigma^2_{X_P})$$

and

(5.45)

$$\sigma^2_{K_\gamma} = \frac{3}{2}\cdot(\sigma^2_{X_N}+\sigma^2_{X_P})$$

and the means are simply $\mu_{\omega_{o,\gamma}/\omega_o} = \mu_{Q_{o,\gamma}/Q} = \mu_{K_\gamma} = 1$.

These are very simple but useful results. They effectively relate the variances of the final biquad deviations to that of I_S. For instance, if we know (say, from the silicon foundry) that $\sigma^2_{X_N} = \sigma^2_{X_P} = \sigma^2_X = 0.01$ (i.e., 95% of the time[†] we will find $I_{S,\,actual}/I_{S,\,nominal} \in [0.80, 1.20]$), the variances of $\omega_{o,\gamma}/\omega_o$ (or Q_γ/Q) and K_γ will be 0.005 and 0.03 respectively. In physical terms, if the desired biquad has $f_o = $ 1MHz, $Q= 1$, and $K= 1$, we can predict that, due to area mismatches, 95% of the fabricated filters will exhibit: $f_o \in [0.86\text{MHz}, 1.14\text{MHz}]$, $Q \in [0.86, 1.14]$, and $K \in [0.66, 1.34]$ [\].

We have performed a Monte Carlo analysis on the log-domain biquad filter. Random variations are added to the I_S value of each transistor according to (5.42). SPICE simulations are repeated 1000 times, so that 1000 simulated samples of biquad parameter's f_o, Q and K are collected. The statistical mean and variance of these parameters are computed. Three values of σ^2_X are chosen for the analysis. The statis-

---

†. If X is a Gaussian r.v., the 95% confidence interval is defined as:

$$P(\mu_X-\Delta X < X < \mu_X+\Delta X) = 0.95 \qquad \text{where} \qquad \sigma^2_X = \left(\frac{\Delta X}{1.96}\right)^2$$

\. We are assuming the resulting filter deviations are also Gaussian distributed, so that we can conveniently employ the 95% interval. Fortunately, as we apply the *Goodness of Fit Test* [62] to the Monte Carlo simulated data to be shown later, this assumption is indeed found to be valid (see Appendix E).

tical filter response deviations when $\sigma_X^2 = 0.001$ is displayed in Figure 5-14, which closely approximate a Gaussian distribution. The simulated results are summarized in Table 5-4, where one can compare these results to the theoretical predictions given in (5.45) and find that they are in very good agreement. In addition, it is also evident that the accuracy will improve with decreasing σ_X^2, which is a property of the transmission of moment's formula mentioned before.

Before we leave this section, we would point out that area mismatches are caused by local variations in junction doping and in photolithographic delineation of the emitter opening. Besides manufacturing variations, mismatches can also arise due to thermal gradient across the chip. The junction voltage V_{BE} can vary by 2mV per $^\circ$C of temperature difference, which corresponds to an area ratio of approximately 1.08. Therefore, to minimize the nonideal effect, it is important to ensure highly symmetrical layout, such as arranging the four-transistor log-domain cells as cross-connected quads arranged in a common-centroid geometry.

5.5 Summary

Simple formulae that account for log-domain integrator errors due to major transistor nonidealities (such as parasitic emitter and base resistances, finite β, Early voltages, and area mismatches) were derived. They can be used to predict the frequency response deviations experienced by lowpass or bandpass biquadratic filters. Simplifying assumptions were made to tackle the log/exp transcendental nature. Nevertheless, the resulting predictions are practical, and their accuracy have been confirmed by SPICE. Finally, several simple electronic compensation schemes were proposed to correct for these transistor nonidealities.

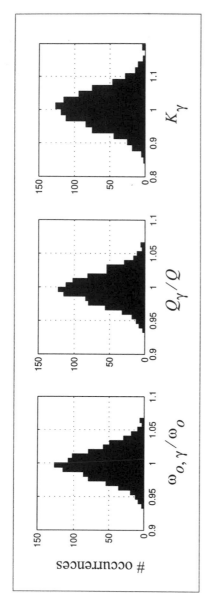

Figure 5-14: Filter response deviations of the log-domain biquad when the variance of area mismatch equals 0.001.

σ_X^2	cutoff frequency		Q factor		filter gain K	
	μ_{f_{oa}/f_o}	σ_{f_{oa}/f_o}^2	$\mu_{Q_a/Q}$	$\sigma_{Q_a/Q}^2$	μ_K	σ_K^2
1e-4	0.9998	5.16e-5	0.9994	5.14e-5	1.0010	2.95e-4
1e-3	0.9988	5.18e-4	0.9985	5.17e-4	1.0040	3.12e-3
1e-2	0.9994	5.31e-3	0.9990	5.31e-3	1.0198	3.32e-2

Table 5-4: Monte Carlo simulations showing effects of area mismatches.

CHAPTER 6

Extending the Nonideality Analysis to High-Order Ladder Filters

The effects of transistor nonidealities on the frequency response of a biquadratic log-domain filter were quantified in the last chapter. A two tier approach was used. First, the effects of transistor nonidealities were related to errors in the integrator circuits, and these, in turn, were related to deviations in the filter's frequency response. In addition, a simple circuit technique to compensate for their effects was described. SPICE simulations confirmed the practicality and accuracy of the theory. In this chapter, we will extend this theory to include the effect of transistor nonidealities on high-order log-domain filters synthesized from an LC ladder prototype.

6.1 Relationship to Classical Filter Theory

In the previous chapter, the effects of transistor nonidealities on integrator operation was described with the help of two error terms: *scalar error (*denoted by *k)* and *parasitic feedback error (*denoted by *f)*. These terms arose quite naturally as they correspond directly with integrator bias current error and the residual dc current alongside the integrating capacitor. Nonetheless, these errors can also be related to another set of error terms known as integrator *magnitude* $m(\omega)$ and *phase* $\theta(\omega)$ *errors*. These are often used to describe the nonideal operation of linear integrators. Assuming that these errors are small, $m(\omega)$, $\theta(\omega) \ll 1$, a nonideal linear integrator can be expressed in the form [3]

$$T_{nonideal}(\omega) = \frac{1}{j\omega\tau_{int}} \cdot (1 + m(\omega) + j\theta(\omega)) \tag{6.1}$$

where τ_{int} is the ideal integrator time constant. Since biquadratic filters are formed by connecting two integrators (namely T_1, T_2) in a feedback loop, it is well known that their actual response, measured in terms of ω_{oa} and Q_a, are related to the integrator errors as follows (assuming high Q),

$$\frac{\omega_{oa}}{\omega_o} = 1 + \frac{1}{2}(m_1(\omega_o) + m_2(\omega_o))$$

and

$$\frac{Q_a}{Q} = \frac{\omega_{oa}/\omega_o}{1 + [\theta_1(\omega_o) + \theta_2(\omega_o)] \cdot Q} \tag{6.2}$$

In contrast, the transfer function of a log-domain integrator suffering from a scalar error (k) can be described as

$$T_{nonideal}(\omega) = \frac{k}{j\omega\tau_{int}} \tag{6.3}$$

while one with a feedback error (f) becomes

$$\begin{aligned}T_{nonideal}(\omega) &= \frac{1}{j\omega\tau_{int}} \cdot \frac{1}{1 + f/(j\omega\tau_{int})} \\ &\approx \frac{1}{j\omega\tau_{int}} \cdot \left(1 + j\frac{f}{\omega\tau_{int}}\right)\end{aligned} \tag{6.4}$$

Comparing (6.3), (6.4) with (6.1), we conclude that the log-domain integrator magnitude and phase errors are, respectively, given by[†]

$$m(\omega) = k - 1 \qquad \theta(\omega) = \frac{f}{\omega\tau_{int}} \tag{6.5}$$

Equation (6.5) relates the concept of integrator errors (magnitude and phase) to our particular description of log-domain integrator errors (scalar and parasitic feedback). Substituting (6.5) into (6.2), and knowing that both log-domain integrators

---

†. The fact that the integrator phase error is inversely proportional to frequency may seem peculiar. However, this is indeed common to non-closed loop structures, such as the Gm-C filters [63]. And log-domain filters, which are also essentially open-loop, are no exception.

have equal k and f, the biquad filter response deviations can be evaluated to be

$$\frac{\omega_{oa}}{\omega_o} = k$$

and

$$\frac{Q_a}{Q} = \frac{k}{1 + \left[\dfrac{f}{\omega_o \tau_{int1}} + \dfrac{f}{\omega_o \tau_{int2}}\right]Q}$$

$$= \frac{k}{1 + \left[fQ + \dfrac{f}{Q}\right]Q} \qquad\qquad (6.6)$$

$$\approx \frac{k}{1 + fQ^2}$$

Interesting enough, (6.6) is identical to the expressions derived previously in Chapter 5 for the lowpass biquad, i.e. (5.23). Therefore, just like other conventional filtering techniques, log-domain filter nonidealities can be investigated using the concept of integrator magnitude and phase errors. Moreover, the well-established theories of nonideal biquad operation (such as (6.1)-(6.2)) can be applied to the relatively new log-domain filtering technique.

The nonideal log-domain integrator magnitude and phase errors are computed using (6.5) and tabulated in Table 6-1 for four transistor nonideal parameters, namely the emitter resistances, finite beta, base resistances and Early voltage.

Unlike the integrator errors for other technology (say, active RC), no distinction has been made between errors of log-domain inverting or noninverting integrators. For simplicity, we have always analyzed the nonideal log-domain positive and negative integrators in pairs, and they have not been studied alone. Judged from the results for the biquadratic filters in Chapter 5, this approach is reasonable. Simple equations with accurate results are achieved. The high-order filter nonideality calculations to be presented shortly will further verify this claim.

6.2 Study of Nonideal High-Order Log-Domain Filters

High-order log-domain filters synthesized using the method of operational simulation of LC ladders have a direct correspondence to their passive LC ladder prototypes. Consequently, much insight into nonideal filter behavior can be obtain through the application of LC ladder network theory. In this section, we will review some of the important LC ladder sensitivity formulae, and then demonstrate how log-domain circuits can benefit from them.

Transistor Nonidealities	integrator magnitude error $m(\omega)$	integrator phase error $\theta(\omega)$	Notes
RE	$\dfrac{-R_E I_o}{V_T + R_E I_o}$	0	cf. Eqn. (5.13)
β	$\dfrac{2}{\beta+2}$	$\dfrac{1}{\omega\tau_{int}}\left(\dfrac{2}{\beta+7}\right)^2$	cf. Eqn. (5.24) see note[a]
RB	$\dfrac{-R_B I_o/\beta}{V_T + R_B I_o/\beta}$	0	cf. Eqn. (5.26)
V_A	$\sqrt{\dfrac{1+\dfrac{V_{CC}-V_{BE}}{V_A}}{1+\dfrac{V_{CC}+V_{BE}}{V_A}}}-1$	$\dfrac{1}{\omega\tau_{int}}\cdot\dfrac{15}{(V_A+50)^2}$	cf. Eqn (5.33)

Table 6-1: Magnitude and phase errors of log-domain integrators.

a. As will be described shortly, β-induced parasitic feedback (or phase) errors are topology dependent. The phase error expression listed for β should be employed with care. For example, the integrator phase error for the high-order lowpass ladder filter due to β will be $\dfrac{1}{\omega\tau_{int}}\left(\dfrac{2}{\beta+2}\right)$.

6.2.1   Nonideal LC Ladder Behavior

LC ladders are probably one of the most thoroughly studied class of circuits known to electronic engineers. Many well-proven theories exist that quantify the filter deviation based on real inductor and capacitor components. In this subsection, we will review some of the most important results [3], [64]-[65], paving the way for their application to log-domain filters.

In most general terms, the deviation of an *LC* ladder transfer function, or more correctly the loss function $H(j\omega) = V_{in}/V_{out}(j\omega)$, to small changes in component values can be approximated by

$$\frac{\Delta H}{H}(\omega) = S_{R_1}^H(\omega)\frac{\Delta R_1}{R_1} + S_{R_2}^H(\omega)\frac{\Delta R_2}{R_2} + \sum_{\text{all } L} S_{L_K}^H(\omega)\frac{\Delta L_k}{L_k} + \sum_{\text{all } C} S_{C_K}^H(\omega)\frac{\Delta C_k}{C_k} \quad (6.7)$$

where R_1 and R_2 are the terminating resistors. The sensitivity functions denoted by the symbol S are written as

$$S_{R_1}^H(\omega) + S_{R_2}^H(\omega) = -\frac{\rho_1(\omega) + \rho_2(\omega)}{2}$$

$$\sum_{\text{all } L} S_{L_K}^H(\omega) = \frac{1}{2}\left[\omega\frac{d\alpha(\omega)}{d\omega} + j\omega\tau_G(\omega) + \frac{1}{2}(\rho_1(\omega) + \rho_2(\omega))\right] \quad (6.8)$$

$$\sum_{\text{all } C} S_{C_K}^H(\omega) = \frac{1}{2}\left[\omega\frac{d\alpha(\omega)}{d\omega} + j\omega\tau_G(\omega) - \frac{1}{2}(\rho_1(\omega) + \rho_2(\omega))\right]$$

where $\rho_1(\omega)$, $\rho_2(\omega)$ are the front-end and back-end reflection coefficients, $\alpha(\omega)$ is the attenuation in Nepers, and $\tau_G(\omega)$ is the group delay expressed in seconds.

Undoubtedly, the equations presented above are very important results. However, they are too general and cumbersome to be of much practical use to us here. Fortunately, there are some special cases in which (6.7) will take on a more usable form, these are described below.

Uniform Component Tolerances

We assume each element type has an *uniform tolerance* such that

$$\varepsilon_R = \frac{\Delta R_1}{R_1} = \frac{\Delta R_2}{R_2} \qquad \varepsilon_L = \frac{\Delta L_k}{L_k} \qquad \varepsilon_C = \frac{\Delta C_k}{C_k} \qquad \forall k \quad (6.9)$$

Based on (6.7) and (6.8), the attenuation deviation ($\Delta\alpha$, in Nepers) can be derived as

$$\Delta\alpha(\omega) = Re\left[\frac{\Delta H}{H}(\omega)\right]$$

$$= -\varepsilon_R Re\left(\frac{\rho_1 + \rho_2}{2}\right) + \frac{\varepsilon_L + \varepsilon_C}{2}\left[\omega\frac{d\alpha(\omega)}{d\omega}\right] + \frac{\varepsilon_L - \varepsilon_C}{2}\left[\frac{1}{2}Re(\rho_1 \cdot \rho_2)\right] \qquad (6.10)$$

As a special case when $\varepsilon_R = 0$, $\varepsilon_L = \varepsilon_C = \varepsilon$, (6.10) becomes

$$\Delta\alpha(\omega) = \varepsilon\omega\frac{d\alpha(\omega)}{d\omega} \qquad (6.11)$$

This is also the case when the filter is *frequency-scaled*, resulting in a new transfer function: $H'(\omega) = H((1 + \varepsilon) \cdot \omega)$. In addition, it has been shown that an upper bound [64] on $\Delta\alpha(\omega)$ is given by

$$|\Delta\alpha(\omega)|_{max} < \frac{|\varepsilon||\rho(\omega)|\omega\tau_G(\omega)}{1 - |\rho(\omega)|^2} \qquad (6.12)$$

Uniform Component Dissipations

Our next consideration is when the reactive elements have a uniform dissipative effect. We can model this as complex increments of the nominal component values as depicted in Figure 6-1, such that

$$sL_{nonideal} = sL\left(1 + \frac{d_L}{s}\right)$$

$$\frac{1}{sC_{nonideal}} = \frac{1}{sC\left(1 + \frac{d_C}{s}\right)} \qquad (6.13)$$

Assuming a *semi-uniform* dissipation described by

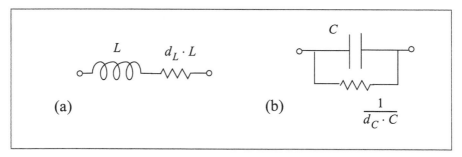

(a)   (b)

Figure 6-1: Models for (a) a dissipative inductor, and (b) a dissipative capacitor.

$$\frac{\Delta L_k}{L_k} = \frac{d_{L_k}}{s} = \frac{d_L}{s} \qquad \frac{\Delta C_k}{C_k} = \frac{d_{C_k}}{s} = \frac{d_C}{s} \qquad \forall k \qquad (6.14)$$

where d_L, d_C are constants, (6.7) and (6.8) will combine to give the filter magnitude deviation as

$$\Delta\alpha(\omega) = Re\left[\frac{\Delta H}{H}(\omega)\right]$$

$$= \left(\frac{d_L + d_C}{2}\right)[\tau_G(\omega)] + \left(\frac{d_L - d_C}{2\omega}\right)\left[\frac{1}{2}Im(\rho_1(\omega) + \rho_2(\omega))\right] \qquad (6.15)$$

6.2.2 Applications to Log-Domain Filters

It is well known that integrator magnitude errors are equivalent to component tolerances in the LC ladder prototype, while phase errors correspond to parasitic (resistive) dissipations [3]. To illustrate this, we will revisit the nonideal inductor and capacitor models and then derive their corresponding integration functions. Referring to Figure 6-2, the integration function realized by the inductor (in terms of physical frequencies) can be written as

$$\frac{1}{j\omega L_{nonideal}} = \frac{1}{j\omega L\left(1 + \varepsilon_L + \dfrac{d_L}{j\omega}\right)}$$

$$\approx \frac{1}{j\omega L} \cdot \left(1 - \varepsilon_L + j\frac{d_L}{\omega}\right) \qquad (6.16)$$

Similarly, the integration function implemented by a nonideal capacitor is

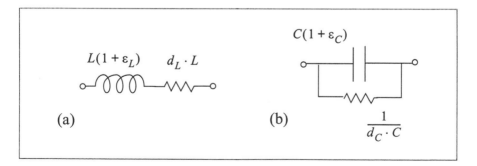

Figure 6-2: Nonideal models for (a) an inductor and (b) a capacitor. It shows both the component tolerance and the parasitic dissipation.

$$\frac{1}{j\omega C_{nonideal}} = \frac{1}{j\omega C\left(1 + \varepsilon_C + \dfrac{d_C}{j\omega}\right)}$$

$$\approx \frac{1}{j\omega C}\cdot\left(1 - \varepsilon_C + j\frac{d_C}{\omega}\right)$$

(6.17)

Comparing (6.16) (or) with the nonideal integration function described in (6.1), we can write

$$m(\omega) = -\varepsilon_L \qquad \theta(\omega) = \frac{d_L}{\omega}$$

(6.18)

or similarly,

$$m(\omega) = -\varepsilon_C \qquad \theta(\omega) = \frac{d_C}{\omega}$$

(6.19)

Therefore, as revealed in (6.18) and (6.19), we observe that the log-domain magnitude and phase errors can be equivalently represented as component drifts and dissipations in the passive ladder network, respectively.

We are now well equipped to apply the *LC* ladder deviation theories directly to the log-domain filter. Equations (6.12) and (6.15) quantify the passive ladder deviations due to component drifts and dissipations. With the log-domain integrators simulating the inductor or capacitor functions, we can equate their errors as follows

$$\varepsilon = -m(\omega) \qquad d_L = d_C = \omega \cdot \theta(\omega)$$

(6.20)

where the expressions for $m(\omega)$ and $\theta(\omega)$ are documented in Table 6-1. We can now apply these results to obtain the attenuation deviation $\Delta A(\omega)$, which is related to $\Delta\alpha(\omega)$ by

$$\Delta A(\omega) = 8.7 \cdot \Delta\alpha(\omega)$$

(6.21)

as expressed in dB. Substituting (6.20), (6.21) into (6.12) and (6.15), the attenuation deviations of log-domain ladder filter due to magnitude and phase errors are found to be

$$\Delta A_{mag}(\omega) < \frac{8.7 \cdot |m(\omega)| \cdot |\rho(\omega)|\omega\tau(\omega)}{1 - |\rho(\omega)|^2}, \text{dB}$$

and

(6.22)

$$\Delta A_{phase}(\omega) = 8.7 \cdot \theta(\omega) \cdot \omega\tau(\omega), \text{dB}$$

respectively. An estimate of the overall filter deviation can then be computed by the combining these two components,

$$\Delta A(\omega) = \Delta A_{mag}(\omega) + \Delta A_{phase}(\omega) \tag{6.23}$$

To gain some appreciation for the magnitude of errors involved and to verify the theory, we shall consider the 7th order lowpass Chebyshev lowpass filter (Figure 3-14 in Chapter 3) as an example. Ideally, it is designed to exhibit passband ripple of 1 dB up to the cutoff frequency of 1 MHz. Analysis of this filter results in[†]:

$$|\rho_{max}(\omega)| = 0.4535 \qquad \tau_G(2\pi \cdot 1\,\text{MHz}) = 4\,\mu s$$

Using (6.22) and (6.23) and the integrator errors tabulated in Table 6-1, we can compute the log-domain filter deviations due to major transistor nonidealities. To begin with, the effects of non-zero RE on the log-domain ladder filter are investigated. The results are summarized in Table 6-2.

Three log-domain filters, each realizing the same transfer function (or simulating the identical *LC* ladder prototype), are realized using different bias currents, namely 20, 100 and 300 μA . They are then subjected to RE nonidealities. The corresponding integrator magnitude errors at the cutoff frequency, $m(2\pi \cdot 1\,\text{MHz})$, are first computed. (Notice from Table 6-1 RE does not cause any integrator phase error, and therefore ΔA_{phase} is excluded from Table 6-2). Next, the filter deviations at 1 MHz are calculated and listed in the table beside the simulated results. As is evident, the calculated values are close to the simulated one. Hence, this study demonstrates the practicality and the validity of the proposed equations.

As discussed before in the previous chapter, the higher the bias current the more severe the nonideal effects of RE. With no surprise, as demonstrated in Table 6-2, this also holds true for high-order structures. Our calculations can clearly capture this trend, and agrees reasonably well with the simulated results.

Next, we will study the effects of finite β on the same filter, in which *both* the integrator magnitude and phase errors ($m(2\pi \cdot 1\,\text{MHz})$ and $\theta(2\pi \cdot 1\,\text{MHz})$) are involved in finding the filter deviation. The results are presented in Table 6-3.

---

†. The maximum reflection coefficient $|\rho_{max}|$ is computed from the passband ripple A_p in dB according to,

$$|\rho_{max}| = \sqrt{1 - 10^{\frac{-A_p}{10}}} \tag{6.46}$$

And the factor τ_G denotes the well-known group-delay expressed in seconds. It is the negative derivative of the filter phase transfer function.

Bias current I_o	RE	integrator m error	Calculated $\Delta A(1\,\text{MHz})$	Simulated $\Delta A(1\,\text{MHz})$
	5 Ω	-0.0039	0.49 dB	0.44 dB
20 μA	10 Ω	-0.0077	0.96 dB	0.87 dB
	20 Ω	-0.0153	1.91 dB	1.86 dB
	5 Ω	-0.019	2.37 dB	2.42 dB
100 μA	10 Ω	-0.037	4.62 dB	5.43 dB
	20 Ω	-0.072	8.99 dB	11.31 dB
	5 Ω	-0.055	6.87 dB	8.47 dB
300 μA	10 Ω	-0.105	13.11 dB	16.36 dB
	20 Ω	-0.189	23.59 dB	27.93 dB

Table 6-2: Effects of RE on the high-order lowpass log-domain filter.

As discussed before, finite β affects the filter cutoff frequency as well as the filter shape (for instance, loss of passband ripple). The latter is a manifestation of the integrator phase error. Table 6-3 shows the calculated integrator magnitude and phase errors, followed by the filter deviations caused by them, respectively. The total deviation is achieved by summing the two. Comparing the predicted (calculated) deviations with the simulation results, very good agreement is observed. This again supports the

β	m error	ΔA_{mag}	θ error	ΔA_{phase}	Calculated $\Delta A(1\,\text{MHz})$	Simulated $\Delta A(1\,\text{MHz})$
200	-0.010	1.25 dB	0.0032	0.70 dB	1.95 dB	2.01 dB
100	-0.020	2.50 dB	0.0064	1.40 dB	3.90 dB	3.95 dB
80	-0.024	3.00 dB	0.0079	1.73 dB	4.73 dB	4.88 dB
60	-0.032	3.99 dB	0.0105	2.30 dB	6.29 dB	6.35 dB
40	-0.048	5.99 dB	0.0155	3.39 dB	9.38 dB	9.05 dB

Table 6-3: Effects of beta on the high-order lowpass log-domain filter

soundness of our nonideality analysis on high-order ladder filters.

Finally, to determine the combined effects of the transistor parameters, we can employ the sum of differential's method to find the collective magnitude and phase errors:

$$m(\omega) \approx \frac{\partial m(\omega)}{\partial R_E} \cdot R_E + \frac{\partial m(\omega)}{\partial R_B} \cdot R_B + \frac{\partial m(\omega)}{\partial (1/\beta)} \cdot \left(\frac{1}{\beta}\right) + \frac{\partial m(\omega)}{\partial (1/V_A)} \cdot \left(\frac{1}{V_A}\right)$$

$$\theta(\omega) \approx \frac{\partial \theta(\omega)}{\partial (1/\beta)} \cdot \left(\frac{1}{\beta}\right) + \frac{\partial \theta(\omega)}{\partial (1/V_A)} \cdot \left(\frac{1}{V_A}\right) \tag{6.24}$$

together with the total filter deviation equations given in (6.22)-(6.23). Applying the theory on the same 7th-order log-domain at two different bias currents (100 μA and 10 μA), Table 6-4 summarizes the filter deviation calculations. Again, our predicted deviations agree closely to the SPICE simulated values. Notice that as the bias current increases, the filter will deviate more severely from its ideal response mainly due to the increase of integrator magnitude errors.

In summary, we have demonstrated how we can apply the classical *LC* ladder theories to the log-domain regime using the concept of integrator magnitude and phase errors. Comparisons between hand calculations and the SPICE simulations confirm the formulae validity.

6.3   Compensation of High-Order Log-Domain Filters

Log-domain biquadratic filters can be compensated by (i) tuning the bias current and (ii) injecting a dc current into the integration nodes, as discussed in Chapter 5. This scheme is also directly applicable to the high-order log-domain ladder filters. To facilitate the discussion of the high-order filter compensation, it would be helpful to graphically illustrate where the errors come from, and what effects they have on the final filter shape. Towards that end, an *equivalent passive ladder* is proposed as a visual aid to model the nonideal log-domain filters. Our intention is to tackle the relatively unfamiliar log-domain circuits based on our understanding of *LC* ladders. As will be seen, this one-to-one correspondence will make compensating the log-domain filters a straightforward task.

6.3.1   Parasitic Emitter Resistances (RE)

As demonstrated previously, RE effectively introduces a scalar (magnitude) error to the ideal log-domain integration. For ease of reference, (5.12) is re-captured below,

$$EXP(\hat{V}_o) = \frac{I_o}{2V_T} \cdot \left(\frac{1}{C/k_{RE}}\right) \cdot \int \{EXP(\hat{V}_{ip}) - EXP(\hat{V}_{in})\} dt \tag{6.25}$$

Bias current	Tx models[a]	m	ΔA_{mag} dB	θ	ΔA_{phase} dB	Calculated $\Delta A(1\text{MHz})$ dB	Simulated $\Delta A(1\text{MHz})$ dB
10 μA	i	-0.027	3.38	0.007	1.42	4.81	4.74
	ii	-0.043	5.35	0.008	1.76	7.11	7.04
	iii	-0.077	9.61	0.012	2.72	12.32	11.86
100 μA	i	-0.049	6.13	0.007	1.42	7.55	7.85
	ii	-0.142	17.7	0.008	1.76	19.44	20.53
	iii	-0.280	34.9	0.012	2.72	37.66	36.18

Table 6-4: Combined effects of device nonidealities

a. The three models are the same as those shown in Table 5-1 on page 146.

where k_{RE} is given in (5.13). Comparing with the ideal integrator function (2.5), the effects of RE can be equivalently interpreted as a component drift (where the l and c are changed to l/k_{RE} and c/k_{RE}). This is demonstrated in Figure 6-3.

It is well known that a uniform change to the reactive elements leads to a *frequency scaling* effect. Consequently, the filter cutoff frequency will be changed from f_o to $k_{RE} \cdot f_o$, while the filter shape, such as the passband ripples, remains unchanged. To compensate for its effect, the bias current should be tuned from I_o to I_{comp} (which effectively scales the integrator time constant) according to (5.16). Compensation of the 7th-order ladder filter is demonstrated in Figure 6-4. SPICE simulations are presented in Figure 6-5 to confirm the scheme. It shows the deviations due to RE (under different bias currents) and the compensated response. In both cases, the compensated response corresponds exactly with the ideal response.

Finally, we would like to point out that the above discussion is also applicable to the deviations caused by parasitic base resistances (RB), whereas I_{comp} is now given by (5.28).

6.3.2 Finite Beta

In addition to integrator magnitude errors, finite β also gives rise to phase errors. Phase errors can be associated with a parasitic dissipation in the *LC* ladder. It is common practice [3] to model such dissipations with a small series resistance r_l in series with each inductor and a resistor r_c in parallel with each capacitor as shown in Figure 6-6. Here, they will be employed to represent the effects of the log-domain integrator parasitic feedback (phase) errors associated with finite transistor β.

As described previously in (5.21), the effects of β on the log-domain inte-

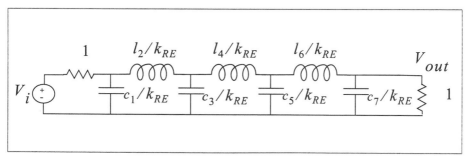

Figure 6-3: Passive ladder equivalence of a 7th-order Chebyshev low-pass log-domain filter under nonzero RE.

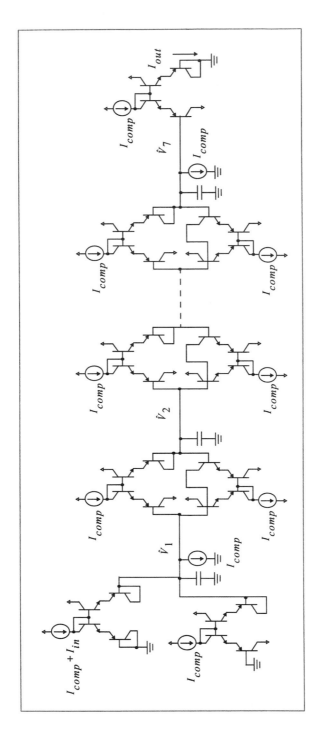

Figure 6-4: Compensation of the nonideal effects due to nonzero RE.

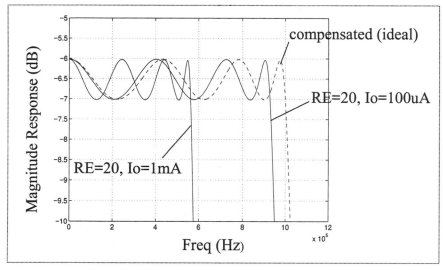

Figure 6-5: Compensation of the 7th-order log-domain filter under finite RE.

gration can be approximated by two factors: k_β and f_β. This equation is re-captured below,

$$EXP(\hat{V}_o) = \frac{I_o}{2V_T} \cdot \frac{1}{C/k_\beta} \cdot \int \left\{ EXP(\hat{V}_{ip}) - EXP(\hat{V}_{in}) - \frac{f_\beta}{k_\beta} \cdot EXP(\hat{V}_o) \right\} dt \quad (6.26)$$

On the other hand, the circuit equations for the lossy inductors and capacitors are written as (with signal symbols defined in Figure 6-6)

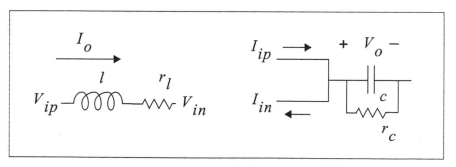

Figure 6-6: Lossy inductor (*l*) and capacitor (*c*).

$$I_o = \frac{1}{l}\int \{V_{ip} - V_{in} - r_l \cdot I_o\} dt$$

$$V_o = \frac{1}{c}\int \left\{I_{ip} - I_{in} - \frac{1}{r_c} \cdot V_o\right\} dt$$

(6.27)

Comparing (6.27) to (6.26), it can be concluded that the log-domain integrator parasitic feedback error (f) can be equivalently modeled by dissipative resistances in the passive LC ladder as

$$r_l = \frac{f_\beta}{k_\beta} \qquad r_c = \frac{k_\beta}{f_\beta}$$

(6.28)

when the following signal correspondences are made,

$$EXP(\hat{V}_{ip}) \Leftrightarrow V_{ip} \qquad EXP(\hat{V}_{in}) \Leftrightarrow V_{in} \qquad EXP(\hat{V}_o) \Leftrightarrow I_o$$

$$EXP(\hat{V}_{ip}) \Leftrightarrow I_{ip} \qquad EXP(\hat{V}_{in}) \Leftrightarrow I_{in} \qquad EXP(\hat{V}_o) \Leftrightarrow V_o$$

(6.29)

In addition, we see that the LC component values are changed to $1/k_\beta$ and c/k_β, respectively. The variables k_β and f_β for the high-order Chebyshev ladder circuit are approximated by[†]

$$k_\beta = \frac{\beta}{\beta + 2} \qquad f_\beta = \frac{2}{\beta + 2}$$

(6.30)

The passive ladder prototype modeling the nonideal effects of finite beta is shown in Figure 6-7. We add a gain factor K_{dc} to complete the nonideal β passive prototype, which is given by[‡]

$$K_{dc} = \left(\frac{\beta}{\beta + 2}\right)^{1 - N/2}$$

(6.31)

Therefore, β error compensation can be achieved electronically by bias current tuning (which is equivalent to correcting the component shift in passive LC ladders) and parasitic feedback cancellation (i.e., removal of dissipative components in passive LC ladders) as shown in Figure 6-8. The filter dc gain is controlled by altering the bias current of the output V-I converter.

Simulations of the 7th-order log-domain ladder filter when $\beta = 80$ and 50 are shown in Figure 6-9. Shift of the cutoff frequency and loss of passband ripple are evident in the plot. When compensation is applied, the magnitude response is nearly restored to its ideal position. Some error near the passband edge is still present, but is

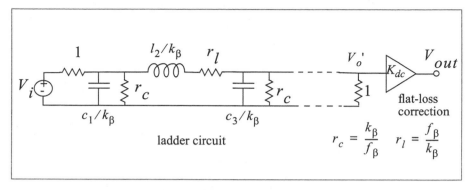

Figure 6-7: Passive ladder equivalence of log-domain filter under finite beta.

at most 0.25 dB.

---

†. This f_β variable is bigger than the one presented in the biquad case (5.24). The nonideal feedback errors due to β are strongly topology-dependent. The more parasitic feedback items (denoted by a [bracket] below) on the KCL equation at the undamped integrating node, the higher the f_β value. For the log-domain biquad (Figure 3-7), at node $\hat{V}_1$ we can write (taking the base current from the subsequent stage into account)

$$\frac{(I_o + I_{in})e^{-\hat{V}_1/(2V_T)}}{\left[1 + \frac{1}{(\beta+1)}e^{-\hat{V}_1/(2V_T)}\right]} - \left[\frac{\beta}{\beta+1}\right]\frac{I_o e^{(\hat{V}_2 - \hat{V}_1)/(2V_T)}}{\left[1 + \frac{1}{(\beta+1)}e^{(\hat{V}_2 - \hat{V}_1)/(2V_T)}\right]} = C_1 \frac{d\hat{V}_1}{dt}$$

For the log-domain 7th-order Chebyshev filter (Figure 3-14(b)), at node $\hat{V}_3$ for instance, we see

$$\frac{I_o e^{(\hat{V}_2 - \hat{V}_3)/(2V_T)}}{\left[1 + \frac{1}{\beta+1}e^{(\hat{V}_2 - \hat{V}_3)/(2V_T)}\right]} - \left[\frac{\beta}{\beta+1}\right]\frac{I_o e^{(\hat{V}_4 - \hat{V}_3)/(2V_T)}}{\left[1 + \frac{1}{\beta+1}e^{(\hat{V}_4 - \hat{V}_3)/(2V_T)}\right]}$$
$$- \left[\frac{1}{\beta+1}\frac{I_o e^{(\hat{V}_3 - \hat{V}_2)/(2V_T)}}{1 + \frac{1}{(\beta+1)}e^{(\hat{V}_3 - \hat{V}_2)/(2V_T)}}\right] = C_3 \frac{d\hat{V}_3}{dt}$$

Obviously, the second expression suffers more parasitic feedback errors, and therefore the corresponding f_β parameter will be higher. A more systematic discussion will be presented in Section 6.4.

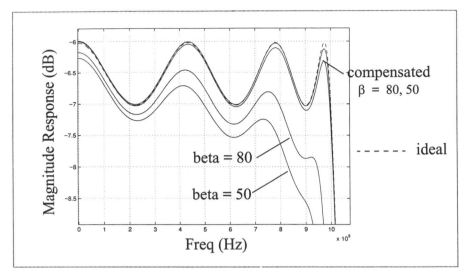

Figure 6-9: Compensations of finite beta on high-order log-domain ladder filter.

6.3.3 Early Voltages

Equipped with the insights developed in section 5.1.4, we can directly draw the nonideal passive ladder equivalent circuit under the influence of finite V_A. Similar to the finite β case, the effects of finite V_A are modeled by drifts in the value of the inductors and capacitors (i.e., l and c become $1/k_{2,V_A}$ and $c/k_{2,V_A}$, respectively) and changes to the resistances (r_l and r_c) according to

---

‡. Without the K_{dc} term, this ladder circuit dictates an extra flat loss at dc due to the stack of parasitic resistances, which roughly equals

$$\left.\frac{V_o{'}}{V_i}\right|_{dc} = -6.02 + \frac{N}{2}\cdot 20\log\!\left(\frac{\beta}{\beta+2}\right)\ \text{dB}$$

where N is the order of the LC ladder filter. However, this expression would be an over-estimate because the parasitic feedback errors of the log-domain filter are all ac signal-dependent, and they should not affect the filter gain at dc. On the other hand, the log-domain ladder filter does suffer a flat loss of $\beta/(\beta+2)$ caused solely by the final EXP V-I converter. Therefore, we append a gain factor K_{dc} to complete the non-ideal β passive prototype.

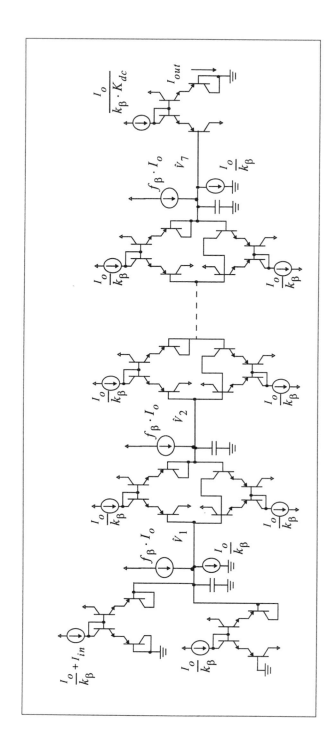

Figure 6-8: Compensation of the nonideal effects due to finite beta.

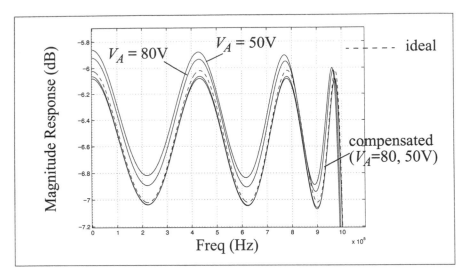

Figure 6-10: Compensations of Early effects on high-order log-domain ladder filter.

$$r_l = \frac{f_{V_A}}{k_{2,V_A}} \qquad r_c = \frac{k_{2,V_A}}{f_{V_A}} \tag{6.32}$$

where k_{2,V_A} and f_{V_A} are given by (5.33). The filter gain factor (K_{dc}) can be approximated by $k_{1,V_A} \cdot k_{2,V_A}$. Compensations can be easily achieved following the schemes presented in the finite β section. SPICE simulation results are presented in Figure 6-10, which demonstrate the effects of the compensation.

6.3.4 Extension to High-Order Bandpass Log-Domain Ladder Filters

Identical to its LP counterpart, the nonideal responses of the ladder BP filter can be characterized by (i) component tolerances and (ii) parasitic dissipations of l and c in the ladder prototype. Taking the 6th-order log-domain Chebyshev BP filter shown in Figure 3-17 as an example, its nonideal passive ladder equivalence is shown in Figure 6-11.

Following the results presented in the previous sections, RE introduces magnitude errors to the log-domain integrators, resulting in center frequency shifting. Referring to the passive ladder, this can be interpreted as having each l and c component value changed to l/k_{RE} and c/k_{RE}, respectively, where k_{RE} is given by (5.13). Besides, as the integrators suffer no phase error, all r_l's and r_c's equal zero. Compen-

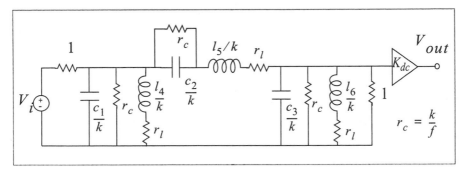

Figure 6-11: Passive ladder equivalence of the nonideal 6th-order Chebyshev log-domain bandpass ladder filter.

sations can of course be achieved through bias current tuning according to (5.16).

Base current causes center frequency shift as well as passband ripple corruption, whose effects can be modeled as in Figure 6-11, where $k \ (= k_\beta)$ and $f \ (= f_\beta)$ are given by $\beta/(\beta + 2)$ and $2/\beta^2$, respectively[†]. Compensation is achieved similar to that described for the high-order LP filter case.

The nonideality analysis developed for Early voltage so far can be directly applied here. With $k \ (= k_{2, V_A})$ and $f \ (= f_{V_A})$ given by (5.33), and gain factor equals $k_{1, V_A} \cdot k_{2, V_A}$. Naturally, the compensation scheme proposed before also holds here.

6.4 Implementation Considerations Under Finite Beta

A closer look at the either filter synthesis method described in this text would reveal that the design is not unique. This stems from the way one utilizes inversion operations. For example, one can choose between a negative integration followed by summation, or a positive integration followed by subtraction. This subtle difference would lead to different log-domain filter implementations, while each of them would implement the same transfer function given ideal devices. However, with

---

†. As we will see in the next section, the parasitic feedback errors (f_β) introduced by finite β are functions of the circuit structure. It can be observed that the BP filter in Figure 3-17 does not enjoy the topological regularity that the LP circuit (Figure 3-14(b)) possesses. Therefore, the f_β factor should be different from node to node.

However, for simplicity, we would assume a uniform f_β for the whole filter, which can be approximated by $2/\beta^2$. Simulation reveals that this is a reasonable approximation.

actual transistors, they may exhibit different levels of deviations. Therefore, it would be advantageous to study the topology-related issues of transistor nonidealities, so that the most robust structure can be selected and designed.

For the transistor parameters studied so far, it is clear that (only) the effects of finite β are topology-sensitive. This is expected because the amount of the undesirable base current is directly determined by the way the integrators are connected, i.e., the circuit topology. To illustrate this point, we would revisit the bandpass and low-pass biquad examples.

By modified nodal analysis, the passive bandpass biquad ladder prototype in Figure 3-8(a) can be completely characterized by the following circuit equations (given that r equals unity),

$$V_o = \frac{1}{c} \cdot \int (V_i - V_o - I_2)dt$$

$$I_2 = \frac{1}{l} \cdot \int V_o dt \qquad (6.33)$$

where V_o/V_i implements the BP biquad transfer function. According to the operational simulation method, the next step would be to convert (6.33) to its SFG representation. Four equivalent SFG's with some minor sign (+/-) differences can be drawn, which will yield four different log-domain filter realizations as shown in Figure 6-12, where $\hat{V}_1$ and $\hat{V}_2$ correspond to V_o and I_2, respectively. Assuming ideal transistors, all of them will implement the same ideal transfer function. However, for finite β, the circuits of Figure 6-12 will experience different effects. To study this, let us write down the KCL equations at the undamped integration node using the equations describing the positive and negative β-corrupted log-domain cell in (5.17) and (5.18). For biquads I and IV of Figure 6-12, we have

$$\left[\frac{\beta}{\beta+1}\right] \frac{I_o e^{(\hat{V}_1 - \hat{V}_2)/(2V_T)}}{\left[1 + \frac{1}{(\beta+1)}e^{(\hat{V}_1 - \hat{V}_2)/(2V_T)}\right]}$$

$$- \frac{I_o e^{-\hat{V}_2/(2V_T)}}{\left[1 + \frac{1}{(\beta+1)}e^{-\hat{V}_2/(2V_T)}\right]} = -C_2 \frac{d\hat{V}_2}{dt} \qquad (6.34)$$

where the error terms caused by β are enclosed between square brackets. Similarly, for biquads II and III, we can write

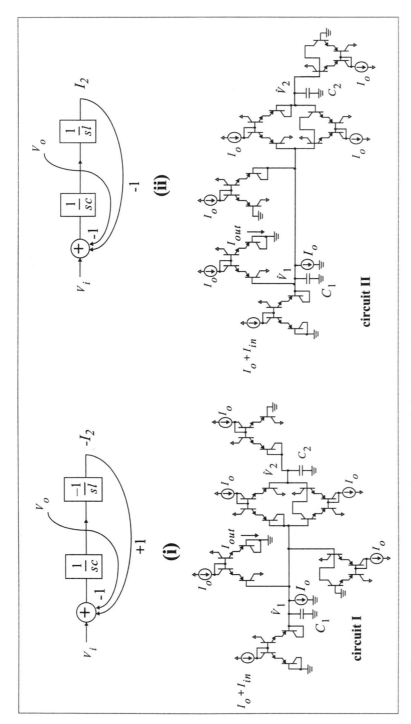

Figure 6-12: Four topologies of the log-domain bandpass biquad filter. *(to be continued on next page)*

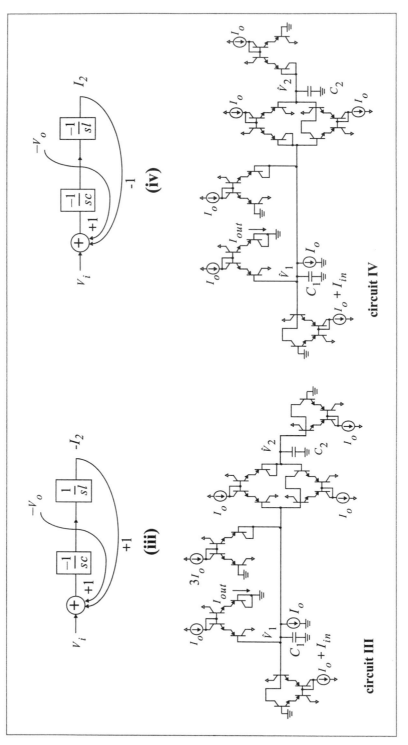

Figure 6-12: *(continued)* **Four topologies of the log-domain bandpass biquad filter.**

$$\frac{I_o e^{(\hat{V}_1 - \hat{V}_2)/2V_T}}{\left[1 + \frac{1}{(\beta + 1)} e^{(\hat{V}_1 - \hat{V}_2)/2V_T}\right]} - \left[\frac{\beta}{\beta + 1}\right] \frac{I_o e^{-\hat{V}_2/(2V_T)}}{\left[1 + \frac{1}{(\beta + 1)} e^{-\hat{V}_2/(2V_T)}\right]}$$

$$- \left[\frac{I_o}{\beta + 1}\right] - \left[\frac{1}{\beta + 1} \cdot \frac{I_o e^{(\hat{V}_2 - \hat{V}_1)/(2V_T)}}{1 + \frac{1}{(\beta + 1)} e^{(\hat{V}_2 - \hat{V}_1)/(2V_T)}}\right] = C_2 \frac{d\hat{V}_2}{dt} \tag{6.35}$$

It is evident that (6.35) has more nonideal terms than (6.34), which, in terms of our analysis method presented in previous sections, dictates a higher parasitic feedback factor f_β. Therefore, under the same β, filters II and III will deviate more severely than I and IV. Due to Q-degradation, their differences are even more prominent when Q is high. A comparison of the biquad filter responses (with f_o=1 MHz, Q=5) with a β =100 is shown in Figure 6-13. The robustness of circuits I and IV over II and III is most evident.

Similarly, there exist four different log-domain filter implementations for the

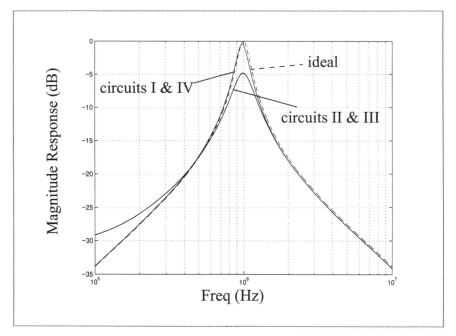

Figure 6-13: Simulated frequency responses showing the effects of beta under different topologies (a) bandpass biquad, (b) lowpass biquad.

lowpass biquad. In increasing order of nonidealities, they are drawn in Figure 6-14(a)-(d) together with their corresponding SFG's. The nonideal integration functions of circuits A to D can then be routinely derived as:

$$
\frac{(I_o + I_{in})e^{-\hat{V}_1/(2V_T)}}{\left[1 + \dfrac{1}{\beta+1}e^{-\hat{V}_1/(2V_T)}\right]} - \left[\frac{\beta}{\beta+1}\right]\frac{I_o e^{(\hat{V}_2 - \hat{V}_1)/(2V_T)}}{\left[1 + \dfrac{1}{\beta+1}e^{(\hat{V}_2 - \hat{V}_1)/(2V_T)}\right]} = C_1 \frac{d\hat{V}_1}{dt} \quad (6.36)
$$

$$
-\frac{\left[\dfrac{(\beta+2)I_o - \beta I_{in}}{\beta+1}\right]e^{-\hat{V}_1/(2V_T)}}{\left[1 + \dfrac{1}{\beta+1}e^{-\hat{V}_1/(2V_T)}\right]} + \left[\frac{\beta}{\beta+1}\right]\frac{I_o e^{(\hat{V}_2 - \hat{V}_1)/(2V_T)}}{\left[1 + \dfrac{1}{\beta+1}e^{(\hat{V}_2 - \hat{V}_1)/(2V_T)}\right]}
$$
$$
+ \left[\frac{I_o + I_{in}}{\beta+1}\right] = -C_1 \frac{d\hat{V}_1}{dt} \quad (6.37)
$$

$$
\frac{\left[\dfrac{\beta}{\beta+1}\right](I_o + I_{in})e^{-\hat{V}_1/(2V_T)}}{\left[1 + \dfrac{1}{\beta+1}e^{-\hat{V}_1/(2V_T)}\right]} - \frac{I_o e^{(\hat{V}_2 - \hat{V}_1)/(2V_T)}}{\left[1 + \dfrac{1}{\beta+1}e^{(\hat{V}_2 - \hat{V}_1)/(2V_T)}\right]}
$$
$$
+ \left[\frac{I_o + I_{in}}{\beta+1}\right] + \left[\frac{1}{\beta+1}\frac{I_o e^{(\hat{V}_1 - \hat{V}_2)/(2V_T)}}{1 + \dfrac{1}{\beta+1}e^{(\hat{V}_1 - \hat{V}_2)/(2V_T)}}\right] = -C_1 \frac{d\hat{V}_1}{dt} \quad (6.38)
$$

$$
-\frac{\left(\left[\dfrac{\beta-1}{\beta+1}\right]I_o - I_{in}\right)e^{-\hat{V}_1/(2V_T)}}{\left[1 + \dfrac{1}{\beta+1}e^{-\hat{V}_1/(2V_T)}\right]} + \frac{I_o e^{(\hat{V}_2 - \hat{V}_1)/(2V_T)}}{\left[1 + \dfrac{1}{\beta+1}e^{(\hat{V}_2 - \hat{V}_1)/(2V_T)}\right]}
$$
$$
- \left[\frac{1}{\beta+1}\frac{I_o e^{(\hat{V}_1 - \hat{V}_2)/(2V_T)}}{1 + \dfrac{1}{\beta+1}e^{(\hat{V}_1 - \hat{V}_2)/(2V_T)}}\right] - 2\left[\frac{I_o}{\beta+1}\right] = C_1 \frac{d\hat{V}_1}{dt} \quad (6.39)
$$

From (6.36) to (6.39), increasing nonidealities are observed, which dictate higher filter response deviations as we move from circuits A to D. To confirm this, SPICE simulations are performed on the LP biquads (with f_o=1MHz, Q=5) when β equals 100. The results are plotted in Figure 6-15 illustrating increased deviation in the order just described. This confirms our predictions.

Since log-domain filters are not unique and may deviate differently with

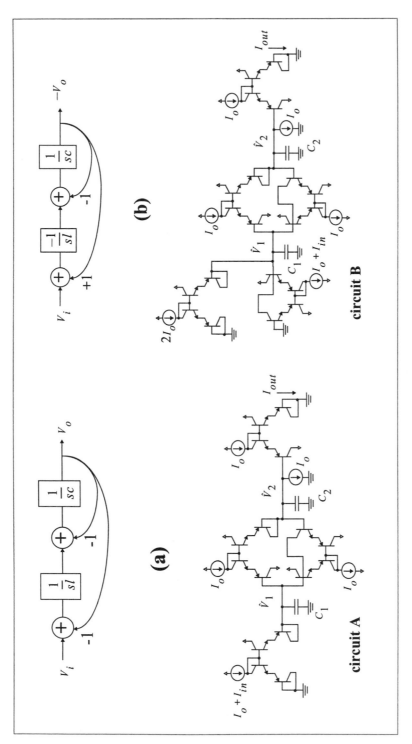

Figure 6-14: Four topologies of the log-domain lowpass biquad filter. (*to be continued on next page*)

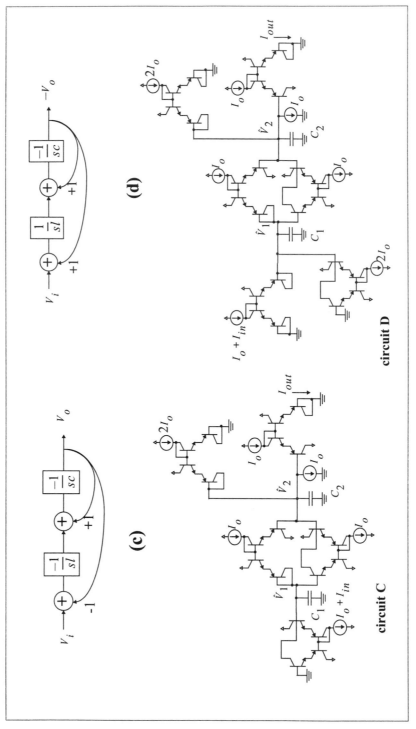

Figure 6-14: *(continued)* Four topologies of the log-domain lowpass biquad filter.

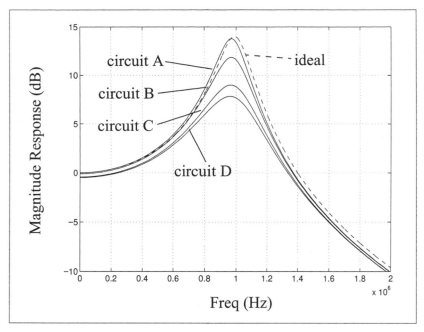

Figure 6-15: Simulated frequency responses showing the effects of beta under different lowpass biquad topologies.

respect to finite β, it is advisable to perform the above analysis during the design process so that the most robust topology is selected. Interesting, we have noticed that the lowpass and the bandpass biquads [22], [31] presented in the literature are among the most β-insensitive structures. Although examples shown here focus on biquads only, the arguments can be extended to cover high-order filters as well.

6.5 Summary

We have extended the log-domain biquadratic nonideality analysis to high-order ladder structures. The important idea behind this study is how the effects of transistor imperfections are collectively described as integrator magnitude and phase errors. By equating the integrator (magnitude and phase) errors to passive ladder (component drift and dissipation) nonidealities, respectively, classical ladder formulae can be readily applied to the log-domain. Qualitatively, we can describe the effects of transistor imperfections on a log-domain ladder-based filter as in Table 6-5. By doing so, we benefit from the vast knowledge already developed for passive LC ladder networks. Moreover, simple quantitative solutions are derived, whose accuracy and practicality are confirmed with SPICE simulations.

Equivalent passive ladder circuits are proposed as a visual aid to better

Transistor Nonidealities	Corresponding log-domain $\int dt$ error	Equivalent effects in a passive LC ladder prototype	Effects on Filter Responses
RE, RB	scalar (magnitude) error	component tolerances	cutoff/ centre freq. shift
β, V_A	parasitic feedback (phase) error	parasitic dissipations	Q-degradation/ loss of pass-band ripples

Table 6-5: Qualitative description of log-domain filter deviations.

understand the deviation mechanisms of high-order log-domain filters. Based on the insights, expressions are derived to quantitatively describe the compensation schemes for the given transistor parameters. Compensating high-order log-domain filters is essentially no different than that employed for biquadratic ones. Simulations of a 7th-order filter example demonstrates the effectiveness of the proposed scheme. It was also shown how this electronic compensation can be applied to the bandpass case.

Following the integrator-based synthesis methods of this text, we have seen how the distribution of sign through the filter structure can give rise to log-domain circuits with different sensitivities to transistor β. Those structures that were least sensitive to β were highlighted.

CHAPTER 7

Experimental IC

Prototypes

To supplement and verify our studies conducted in the previous chapters, we will present several log-domain filter prototypes and their measured performance. In this way we can further explore the practicality and limitations of log-domain filters, and explore issues which simulation failed to reveal.

Table 7-1 shows a quick overview of the seven test chips to be described. They feature lowpass and bandpass biquads, several designs based on the operational simulation of 4th and 5th-order *LC* ladders, and 3rd and 5th-order filters based on state-space formulations.

7.1 Test Chip #1: Lowpass Biquadratic Filter

The first circuit whose experimental performance will be described is the log-domain lowpass biquadratic filter [66] of Figure 3-7. Notice that this biquad is realized using the most basic log-domain integrator of Section 2.1. Although this design is not suited for high-speed or low-power applications, the results presented here are from the very first log-domain filter implementation produced with the techniques described in this text [29]. We shall begin this section with a description of the experimental set-up used for testing the biquadratic filter. The experimental results will then fall into two main categories, namely frequency performance, and linearity and noise measurements. In the first, we compare the actual response of the biquad to its intended transfer function, then show tunability and high-frequency operation. We then present a series of linearity and noise measurements, including total harmonic distortion, intermodulation distortion and signal-to-noise-plus-distortion ratio.

7.1.1   The Test Set-Up

An integrated circuit of the log-domain biquadratic filter was fabricated using Gennum GA911 analog arrays [52]. These semi-custom bipolar arrays are composed of fixed components that can be interconnected in whatever configuration the designer chooses. In other words, the designer has control over a single metal layer that is deposited over a number of fixed components. The Gennum process provides npn and pnp transistors with f_T's of 2.5 GHz and 10 MHz, respectively. A microphotograph of the chip is shown in Figure 7-1.

The capacitors C_1 and C_2 (as in Figure 3-7) were realized off-chip with values of 2.25 nF and 8.49 nF, respectively. The current sources were built using a modified Wilson design. The circuit was biased with supplies of ± 5 V. In order to generate and measure the input and output currents, simple V/I and I/V converters were included in the test circuit as shown in Figure 7-2. The input was generated using an HP3314A function generator while the frequency analysis was performed using an HP3588A spectrum analyzer.

7.1.2   Frequency Performance

The magnitude response of the filter was first measured and plotted in Figure 7-3. The measured response was found by applying a peak-to-peak current input of 100 µA and sweeping it from 10 Hz to 1 MHz. Also shown in the plot is the ideal response. As is evident from the results, the biquad has the expected transfer function. The cutoff frequency is correct within a small experimental error, mainly due to the parasitic junction resistances of the bipolar transistor as described before.

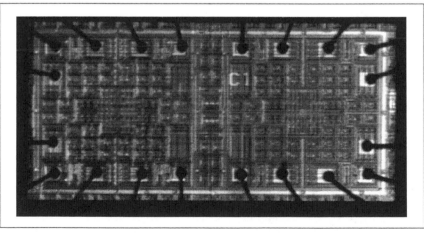

Figure 7-1: Microphotograph of the lowpass log-domain biquadratic filter.

Chip	Filter Type	Signal	Integrator	Process	Synthesis
1	lowpass biquad	single-ended	basic structure (Section 2.1)	Gennum bipolar array	
2	5th-order lowpass Chebyshev		high-speed structure (Section 2.3.1)	Nortel 0.8 μm BiCMOS process	operational simulation of *LC* ladders
3	bandpass biquad	balanced			
4	4th-order bandpass Butterworth				
5	3rd-order lowpass Chebyshev	single-ended	high-speed low-power structure (Section 2.3.3)	Gennum bipolar array	
6	3rd-order lowpass Elliptic (prog.)		high-speed structure (Section 2.3.1)	Nortel 0.8 μm BiCMOS process	state-space
7	5th-order lowpass Elliptic (prog.)	balanced			

Table 7-1: Overview of the seven log-domain filter test chips.

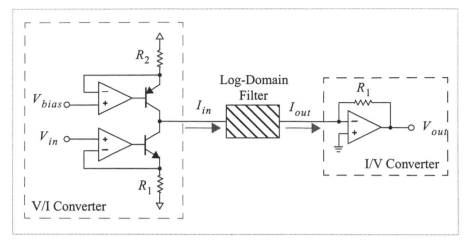

Figure 7-2: The complete test-circuit for the log-domain filter.

One intriguing characteristic of the log-domain filter is its tunability, i.e., the ease in which the cutoff frequency of the circuit can be varied. The tunability of the log filter can be attributed to the factor $I_o/2V_T$ associated with the log-domain integrator. This theory was tested experimentally by varying all of the bias currents in the circuit. The filter showed tunability over two decades, namely from 1 kHz to 100 kHz. A plot of the magnitude responses of the biquadratic filter for three different levels of bias current is shown in Figure 7-4.

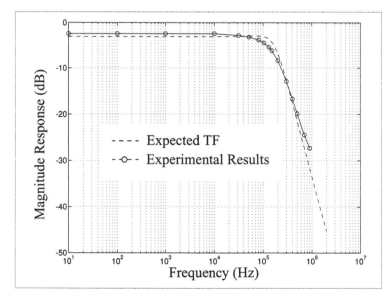

Figure 7-3: Frequency response of the biquad.

The final frequency test, which was performed on the biquadratic filter, was to measure how close the cutoff frequency could be brought to the f_t of the slowest transistor (10 MHz for the GA911 Gennum process). To do so, the capacitors in the circuit were varied such that the cutoff frequency gradually increased. Figure 7-5 shows plots of the magnitude response of the filter for four different cutoff frequencies: 50 kHz, 500 kHz, 1 MHz and 5 MHz. From these results we can see that the passband remains flat up to a cutoff frequency of 1 MHz or 1/10th of the f_t of the slowest transistor.

7.1.3 Linearity And Noise Measurements

One of the most common measures of linearity is total harmonic distortion (THD). This is a measure of the total power of the harmonics to the power of the fundamental. By applying a test signal to the circuit and observing the output on a spectrum analyzer we can measure the harmonics and thus the total harmonic distortion. A spectral plot of the output for the circuit stimulated by a 100 µA, 1 kHz sine wave is shown in Figure 7-6. Given a spectral plot of this kind, the total harmonic distortion can be calculated using

$$\text{THD} = \frac{\sqrt{I_2^2 + I_3^2 + I_4^2 + \dots}}{I_1} \tag{7.1}$$

where I_1 is the current magnitude of the fundamental, and I_2, I_3,... are the current

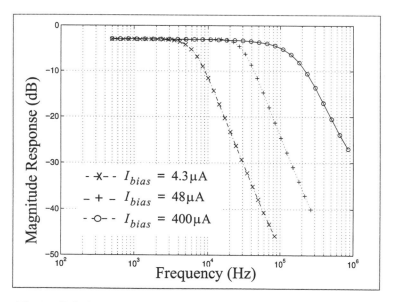

Figure 7-4: Measured tunability of the log-domain biquad.

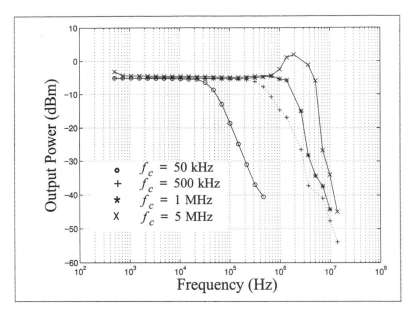

Figure 7-5: Measured magnitude responses of the biquad for four different cutoff frequencies.

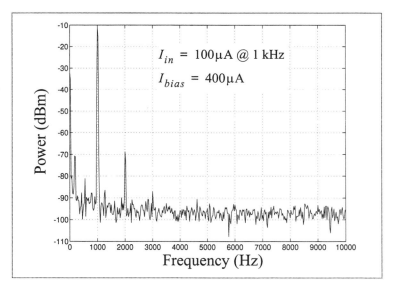

Figure 7-6: Measured spectral plot of the log-domain lowpass biquad.

magnitudes of the harmonics (with the subscript denoting the corresponding harmonic term). Table 7-2 shows THD measurements for a range of different input amplitudes and frequencies. All measurements were taken over a 50 kHz bandwidth. The biquadratic log-domain filter consistently showed distortion levels of less than - 60 dB.

One of the disadvantages of THD as a measure of linearity is that the harmonics may be affected by the filtering characteristics of the device under test. This is particularly relevant when dealing with lowpass or bandpass filters. A better test of linearity in such cases is the measurement of intermodulation distortion (IMD). This involves stimulating a circuit with two tones (at frequencies f_1 and f_2), which are usually close to each other and near the passband edge of the filter. In addition to the usual harmonics present at the output, we expect to see intermodulated harmonics that satisfy the following criteria,

$$f_{nm} = |nf_1 + mf_2| \qquad (7.2)$$

where n and m are positive integers such that $n + m \leq 3$. A spectral plot of the log-domain biquad excited by two tones of slightly less than the cutoff frequency of the filter is shown in Figure 7-7. The components f_{20} and f_{02} represent the second-order distortion products which accounted for most of the total harmonic distortion found in the previous section. The tone f_{11} represents the sum of the two tones and is traditionally 6 dB greater than the second order components, as is the case here. We concentrate here on the distortion components close to the original tones (f_{12} and f_{21}) since these fall near the desired frequencies and can prove the most troublesome [67]. These are known as the third-order intermodulation distortion (denoted as IMD3) products. In Figure 7-8, we show a plot of the sum of the power of these IMD3 products versus the amplitude of one of the input tones. The second tone was of equal amplitude. Also included on the graph is a plot of the power of the sum of the two fundamentals versus the input current level. The least intermodulation distortion occurs for an input of 100 μA and measures -70 dB. The IMD3 products then increase due to the cubic relationship between these harmonics and the input signal. At low input levels, the power of the harmonics remains constant because they are so small that they get lost below the noise floor.

I_{in}	Frequency	THD	
100 μA	1 kHz	-62.4 dB	0.071%
10 μA	1 kHz	-62.9 dB	0.076%
100 μA	5 kHz	-60.3 dB	0.096%

Table 7-2: THD measurements for the log-domain biquad.

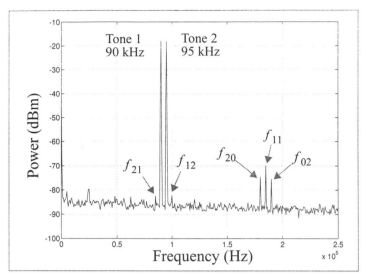

Figure 7-7: Measured IMD of the log-domain lowpass biquad

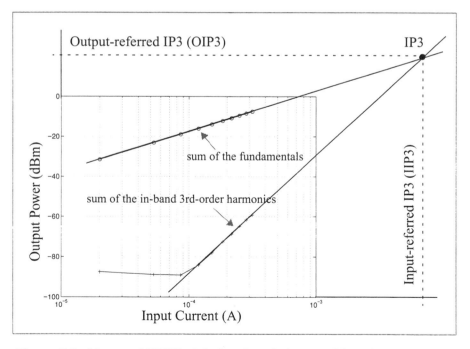

Figure 7-8: Measured IMD3 of the log-domain lowpass biquad versus input current. Also shown is the concept of a third-order intercept point.

As the IMD changes with input signal level, it is common practice to extend a line through the set of measured points as shown in Figure 7-8 and quote the point in which the two lines intersect. In this way, a single measure is used to capture most of the distortion behavior of the circuit, independent of the signal level. In the case of IMD3, this point is called the *third-order intercept point* (IP3*)*. A similar point exists for the second-order IMD, fourth-order IMD, etc. In general, the higher the intercept point the lower the circuit's distortion.

One can derive the input-referred IP3 (denoted IIP3) from a measurement of the circuit's gain G, and any measure of the fundamental output level P_{out1} and the third-order distortion P_{out3} at a given current input level I_{in} expressed in dBs. Specifically as

$$IIP3 = I_{in}\big|_{db} + \frac{P_{out1} - P_{out3}}{2G} \qquad (7.3)$$

For example, to obtain the circuit's gain we measure the change in the output power level of the fundamental ΔP_{out1} to a change in the input signal level ΔI_{in} expressed in dBs, or mathematically as

$$G = \frac{\Delta P_{out1}}{\Delta I_{in}\big|_{db}} \qquad (7.4)$$

For the measured results shown in Figure 7-8, G is found to be 1 dBm/dB. Likewise, for a current input level of 80 μA, P_{out1} = -20 dBm and P_{out3} = -90 dBm, the IIP3 is 4.5 mA (-46.9 dB).

The final test performed on the biquad was to find the signal-to-noise-plus-distortion ratio (SNDR) for different levels of input current. The performance of the biquad biased with 400 μA current sources is shown in Figure 7-9. The measurement was made over a 50 kHz bandwidth. The SNDR increases linearly up to a peak-to-peak signal of 25 μA then we see that the harmonics begin to dominate. The maximum attainable SNDR with this circuit was 54 dB. From these results together with those of the previous sections we see that the distortion behavior of the biquadratic filter is best between 10 and 100 μA. This is consistent with the transistor specifications which shows that the transistors have their highest β levels for currents in that range.

7.2   Test Chip #2: Fifth-Order Chebyshev Lowpass Filter

Our next experiment focuses on a fifth-order Chebyshev log-domain filter [30]. Synthesized using the operational simulation method, the design is similar to the seventh-order example (Figure 3-14(b)) discussed in Section 3.2.3. Again, the basic

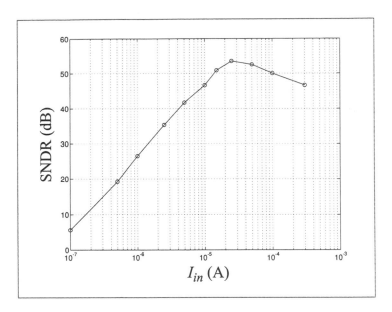

Figure 7-9: Measured SNDR versus the input current (measured over 50 kHz bandwidth).

log-domain integrator of Section 2.1 is employed. In this section, we will show experimental results that verify the performance of this log-domain ladder filter. Again, we focus on the frequency response and linearity of these circuits since these best allow us to judge the usefulness of these new filter techniques.

The fifth-order Chebyshev filter was fabricated with a larger version of the Gennum GA911 bipolar array used for the biquadratic filter. As with the biquad, modified Wilson current mirrors were used for the current sources and the circuit was biased at ±5 V. The external test circuitry and equipment was the same as that described previously in Section 7.1.1.

7.2.1 Frequency Performance

The components of the Chebyshev filter were chosen such that it had a cutoff frequency of 50 kHz and a 1 dB ripple. A plot of the frequency response of the filter is shown in Figure 7-10(a) along with a close-up of the passband (Figure 7-10(b)). The figure also shows a plot of the frequency response of the *LC* ladder which met the original specifications, along with *HSPICE* AC analysis of the filter modeled using both ideal and Gennum transistors.

The results show that the fifth-order log-domain filter has very much the desired frequency response. Most importantly, we see the desired rippling behavior in the passband and the correct attenuation in the transition and stopband regions (until the noise floor is reached). There is a shift in the cutoff frequency by approximately

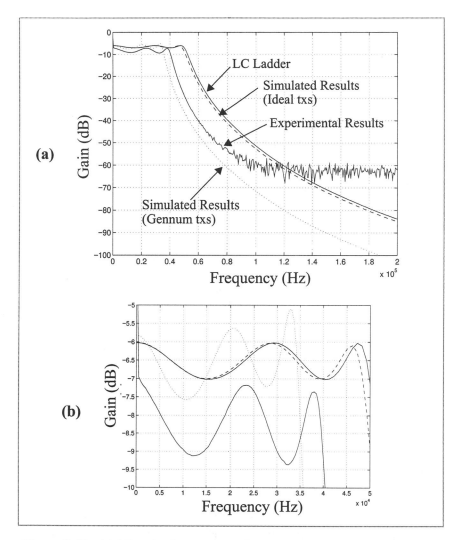

Figure 7-10: (a) Magnitude response of the 5th-order log-domain Chebyshev lowpass filter, and (b) close-up of the passband.

10 kHz that is due to the non-ideal nature of the transistors. More specifically, as discussed in detail before, it is mostly due to the parasitic junction resistances of the bipolar transistor. In addition, the passband ripple is corrupted due to the finite beta of the actual bipolar devices.

The tunability of this filter is demonstrated in Figure 7-11. Although the resistors that set the bias current of the circuit were fixed, we could still vary the supply voltages and thus change the bias in this manner. The general circuit operation was maintained although the new supply levels did have a slight effect on the filter

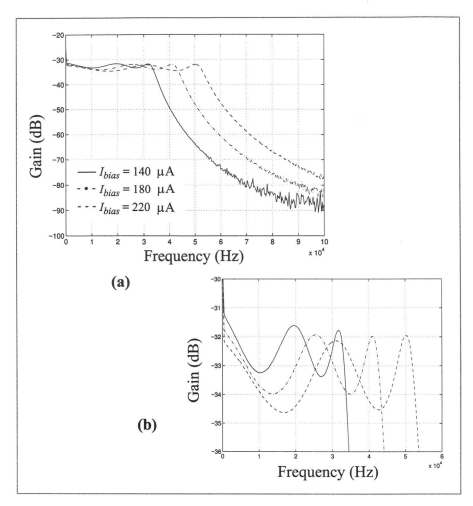

**Figure 7-11: Measured tunability of the 5th-order Log-domain filter:
(a) fully frequency range, and (b) passband.**

response, as will be discussed shortly. Figure 7-11 shows the fifth-order Chebyshev filter biased with ±4V, ±5V and ±6V supplies which corresponded to bias currents of 140 µA, 180 µA and 220 µA respectively. Clearly, we can see how the bias current controls the cutoff frequency of the circuit while maintaining most other filter characteristics. Note however, that the passband ripple increased slightly as the supply voltages were increased. This is due to the Early effect that modifies the ideal exponential nature of the bipolar transistor. It becomes more pronounced when higher voltage supplies are used since it is related to the transistor voltage V_{CE}.

7.2.2 Linearity Measurements

Figure 7-12 shows a plot of the total harmonic distortion of the fifth-order Chebyshev log-domain filter versus input current. The input was a 2 kHz sine wave of varying amplitude. The frequency of the input tone was chosen such that it was well below the cutoff frequency of the filter so that the harmonics were not affected by the natural attenuation of the filter. The fact that the input frequency is below the passband edge leads to slightly better distortion measurements than might be found for an input signal placed right at the cutoff frequency. An intermodulation distortion test is generally a better measure of linearity.

The best harmonic distortion measure was -47 dB found for an input of 15 µA and a bias current of 180 µA. The distortion is primarily due to the first harmonic found at twice the frequency of the fundamental, as shown in a spectral plot of the filter stimulated by a 30 µA, 2 kHz sine wave in Figure 7-13.

In order to measure the intermodulation distortion (IMD) of the log-domain filter a two-tone stimulus was applied to the filter. The frequencies of the two tones were chosen such that they were 4 kHz apart and that they were placed slightly below the passband edge of the filter. Since this is where the worse distortion occurs, this gives us a form of worst-case analysis of the linearity of the filter. Figure 7-14 shows the spectral response of the Chebyshev filter stimulated by tones of 28 kHz and 32 kHz. One can clearly see the different harmonics and intermodulation products.

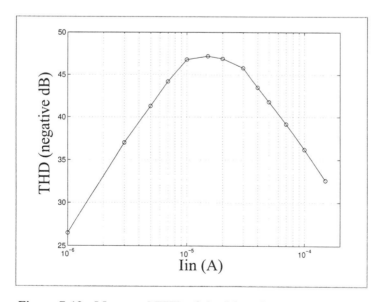

Figure 7-12: Measured THD of the 5th-order log-domain filter versus input current.

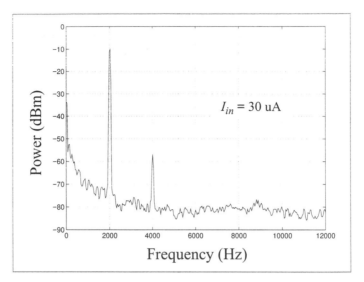

Figure 7-13: Measured spectral plot at the filter output for a single tone input.

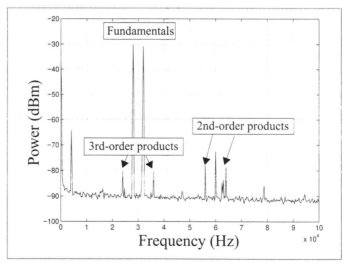

Figure 7-14: Measured spectral plot at the filter output for a two-tone input.

The products that are of the most interest to us are the third-order tones found at 24 kHz and 36 kHz. Figure 7-15 shows a plot of the sum of the power of the fundamentals and the sum of the power of the third-order harmonics versus the input current amplitude. The difference between these two measurements gives us the intermodulation distortion. The best IMD measure for this filter was -55 dB and occurred for an input of 30 μA when biased at 180 μA. The IIP3 metric for this filter is found to be 0.71 mA. For comparison purposes, Figure 7-15 also shows the sum of the power of the second-order products. The distortion measure found by evaluating these products is in effect the same as the THD measurement made in the previous section. Unfortunately, these distortion components are affected by the fact that these products fall in the stopband region of the filter.

7.2.3   Approximating an Elliptic Filter Response

In Section 3.6, we have seen how an approximated elliptic filter response can be realized using a log-domain circuit. Comparing the filter circuit in Figure 3-26 to that of Figure 3-14(b), it can be observed that the elliptic filter bears almost identical structure to its Chebyshev counterpart. In fact, their only differences lie in the value of the integrating capacitors, and in the addition of two bypass capacitors. Therefore, based on the same silicon chip discussed so far, capacitors (including those implementing bypass functions) of appropriate values will be employed to achieve an elliptic response.

The magnitude response of the elliptic log-domain filter is shown in Figure 7-16. The filter was based on an *LC* ladder with a cutoff frequency of 40 kHz and a 1

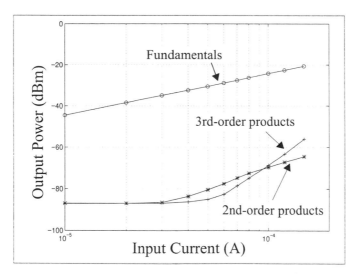

Figure 7-15: Measured IMD of the log-domain filter versus input current.

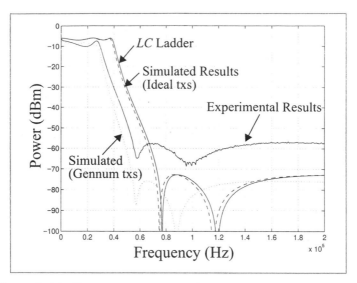

Figure 7-16: Frequency response of the fifth-order elliptic log-domain filter.

dB passband ripple. Upon examining the experimental response of the log-domain filter, we see that the basic response of an elliptic filter has been achieved. We can clearly see the presence of the finite transmission zeros which give rise to stopband ripple. The similarity between the experimental and expected results supports the approximation made in Chapter 3. The linearity of the log-domain elliptic filter was comparable to the results that were found for the fifth-order Chebyshev filter and are not repeated here.

Overall, the log-domain filter experimentations discussed so far showed good correlation between their frequency responses and that required by the specifications. There was some degradation of the passband ripple and a slight shift in cutoff frequency due to the mechanism discussed in Chapters 5 and 6. The filters proved to be tunable over two decades and could be operated up to one tenth of the f_T of the lowest transistor in the circuit. Distortion levels measured using a variety of methods ranged from -45 dB to -70 dB. This is comparable to distortion levels found in many continuous-time filtering schemes [1].

7.3   Test Chip #3: High-Speed Bandpass Biquadratic Filter

In order to assess the potential of log-domain filters for very high frequency (VHF) applications, the bandpass log-domain biquad similar to that shown in Figure 3-11 was realized [42]. However, it is important to note here that the integrators employed are no longer the basic structure of Figure 2-2, but rather the high-speed (all-npn) version of Figure 2-7 in Section 2.3.1.

7.3.1 The Test Set-Up

The log-domain bandpass biquadratic filter was implemented in a 0.8 μm BiCMOS process provided by Northern Telecom having npn transistors with peak f_T of 11 GHz. The current sources connected to the positive supply were realized as high-swing pmos cascode current mirrors, and those connected to the negative power supply were realized as base-compensated cascode current mirrors using npn transistors. A microphotograph of the chip is shown in Figure 7-17. The integrating capacitors, placed at the center of the chip, were 20 pF each. Two input and output on-chip decoupling capacitors are shown at the upper left and right corners. The circuit was packaged in a standard DIP.

In order to keep the circuitry simple and fast for a first prototype, the V/I conversion at the input was implemented using an off-chip 1 kΩ resistor connected to the input current node. This is shown in Figure 7-18 (a). Similarly, the I/V conversion of the output current was performed through a 50 Ω resistor connected to the collector of the output *EXP* transistor (which carries the filter output current), as shown in Figure 7-18(b).

7.3.2 Frequency Performance

Assuming different bias currents for the integrators and the output *EXP*

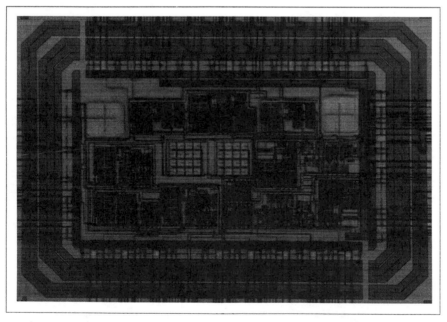

Figure 7-17: Microphotograph of the VHF bandpass log-domain biquad.

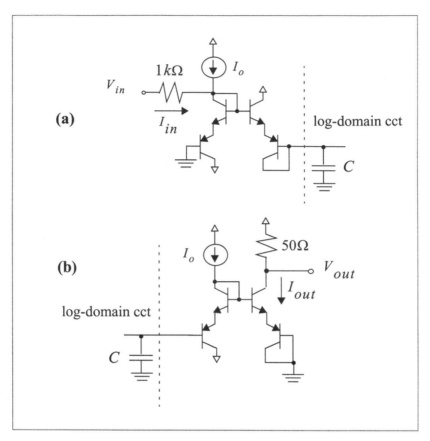

Figure 7-18: Combined voltage-to-current or current-to-voltage converter and log-domain interface circuit: (a) input V-to-I converter and *LOG* operation, (b) *EXP* operation and output I-to-V converter.

circuit, the bandpass biquad transfer function can be explicitly expressed as [42]

$$\frac{I_{out}}{I_{in}}(s) = \frac{K \cdot s}{s^2 + s\left(\dfrac{\omega_o}{Q}\right) + \omega_o^2}$$

$$K = \frac{I_3}{I_1}; \qquad \omega_o = \frac{1}{2V_T} \cdot \sqrt{\frac{I_1 I_2}{C_1 C_2}}; \qquad Q = \sqrt{\frac{I_2}{I_1} \cdot \frac{C_1}{C_2}}$$

(7.5)

where I_1 is the bias current for the damped positive integrator (with capacitor C_1), I_2 is the bias current for the negative integrator (with capacitor C_2), and I_3 is the bias current for the output *EXP* circuit. Therefore, both the filter center frequency and the Q-factor can be electronically tuned and controlled by varying the bias currents.

Figure 7-19 shows frequency tuning from 83.1 MHz to 221.9 MHz (when the bias level changes from 790 µ*A* to 2.2 mA) while maintaining a constant *Q*-factor ($Q \approx 12$). Slight manual adjustments of the biasing current were necessary in order to compensate for the *Q* variations observed when the center frequency was changed. In a practical application, this will be achieved using automatic *Q*-tuning circuitry in order to ensure the stability of the filter.

On the other hand, Figure 7-20 shows the *Q*-factor tuning ($Q \approx 2, 4.1, 12.7$) for a given fixed frequency. A continuous increase in the *Q* up to approximately 50 could be achieved while maintaining stable operation. Higher *Q*'s caused the filter to oscillate.

7.3.3 Linearity And Noise Measurements

A two-tone test was performed to measure IMD3 products of the filter. Figure 7-21 shows a spectrum of the output when two tones are applied at 78 MHz and 83 MHz. Both tones are within the passband of the filter and combine to give a current modulation index (I_{in}/I_{bias}) of *50%* when biased at a current level of 790 µA. The IMD3 measured was -34.7 dB. This corresponds to an IIP3 metric of 2.1 mA. An increase in distortion was observed when the filter's *Q* was raised. For a center frequency of 220 MHz and a modulation index of 50%, the IMD3 was measured to be

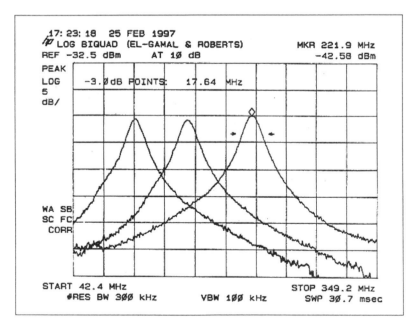

Figure 7-19: Tuning the center frequencies of the bandpass log-domain biquad.

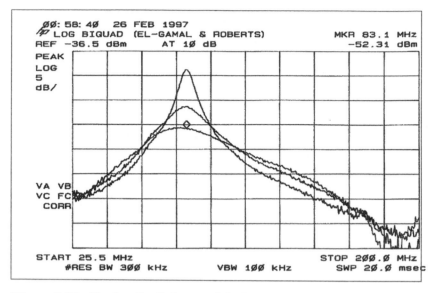

Figure 7-20: Tuning the Q-factor of the bandpass log-domain biquad.

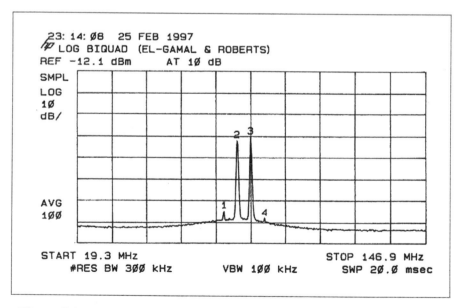

Figure 7-21: Measured intermodulation distortion of the bandpass biquad.

approximately -32.5 dB.

Finally, Figure 7-22 shows the output noise spectrum using a resolution bandwidth of 300 kHz. The input current was set to zero with I_{bias} = 790 µA and Q = 10. The output noise power spectral density was measured to be -123.7 dBm/Hz (146 nVrms/$\sqrt{Hz}$) at the center frequency. The equivalent noise bandwidth of a second-order function was used to compute the output noise power of the filter, which was found to be -52.5 dBm (527 µVrms). The power consumption of the chip was 90 mW - 235 mW for the tuning range of 83 - 220 MHz.

7.4 Test Chip #4: Balanced High-Speed Fourth-Order Bandpass Filter

It is a well-known fact that analog circuits with balanced (or fully-differential) structures are always preferable for reducing a wide range of nonidealities such as noise and distortion. In order to show the feasibility of high-order, high-speed, low-distortion log-domain filters, a fourth-order maximally-flat balanced bandpass filter [42] was designed based on the differential log-domain integrator of Figure 2-8 and synthesized using the method of operational simulation of LC ladder networks discussed in Section 3.2.4. Note that a common-mode feedback (CMFB) circuit is used between each pair of integrating nodes.

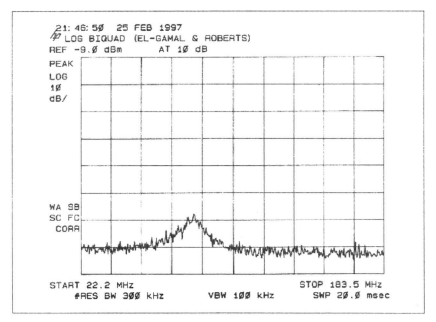

Figure 7-22: Measured biquad's output noise power spectral density.

7.4.1 The Test Set-Up

Practically, a single-ended to differential current-to-voltage and voltage-to-current converter are needed to test the balanced filters. In order to ensure that their performance will not mask the characteristics of the filter under test, those circuits need to meet stringent specifications. They need to be superior to, or at least compatible with, the quality of the filters. Namely, they should be compatible with its operating frequency and with its current signal levels, they should introduce lower distortion than the filter, and provide adequate input and output impedance matching, which is critical at high frequencies. Finally, those interface stages need to be suitable for integration on chip. However, this is not mandatory for characterizing the filters. If available, they widen the scope of applications for log-domain filters to traditional voltage-mode systems.

The single-ended-voltage to differential-current input stage employed is shown in Fig. 7-23. It is based on Caprio's V/I converter [68]-[69]. Transistors Q_1-Q_2 along with their emitter degeneration resistance R convert the input voltage V_{in} into a current I_{in}. Assuming all transistors are matched and neglecting the base currents, the value of I_{in} is given by

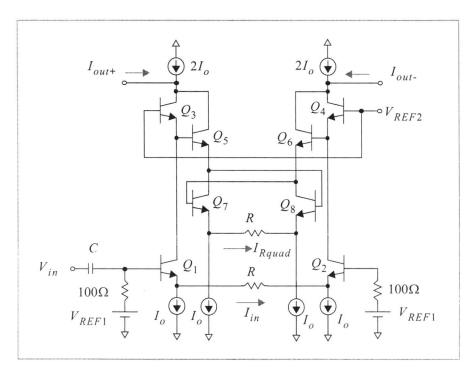

Figure 7-23: A very low-distortion V/I input stage.

$$I_{in} = \frac{V_{in}}{R} - \frac{1}{R} \cdot (V_{BE1} - V_{BE2})$$

$$= \frac{V_{in}}{R} - \frac{1}{R} \cdot \left[V_T \ln\left(\frac{I_o + I_{in}}{I_S}\right) - V_T \ln\left(\frac{I_o - I_{in}}{I_S}\right) \right] \quad (7.6)$$

It is clear from (7.6) that I_{in} will suffer distortion due to the difference in the base-emitter voltages of Q_1 and Q_2. Transistors Q_3 and Q_4 duplicate V_{BE1} and V_{BE2} and apply them, with opposite polarity, to the bases of Q_5 and Q_6. Transistors Q_5- Q_8 constitute a Caprio quad with an input voltage $V_{inquad} = V_{B5} - V_{B6}$. For a small differential input voltage, the current in the resistance of this quad is given by

$$I_{Rquad} = -\frac{V_{inquad}}{R}$$

$$= -\frac{1}{R} \cdot (V_{B5} - V_{B6}) \quad (7.7)$$

$$= \frac{1}{R} \cdot (V_{BE1} - V_{BE2})$$

Adding the current components in (7.6) and (7.7) together results in a theoretically linear output current. Other nonideal effects like the finite gains of the transistors will still cause distortion. However, simulation reveals that the total harmonic distortion of the output current of the V/I converter is substantially better than that of a log-domain biquad, thus suggesting the suitability of the V/I converter of Fig. 7-23 to test the filters. The input is DC decoupled with an on-chip capacitor, and the two 100 Ω resistors ensure a 50 Ω input impedance match. Stability was verified through extensive simulation.

At the output side, the differential filter output voltages are obtained first passing the log-domain voltages through an *EXP* circuit whose collector current sources are replaced by resistors and whose emitter terminal is connected to a reference voltage level as shown in Figure 7-24 [70]. This differential linear voltage is then converted to a single-ended voltage through a circuit similar to that of Fig. 7-23 with the upper current sources replaced by load resistors. Finally, the output voltage is buffered off-chip through a Darlington pair providing a 50 Ω output impedance over the frequency range of interest.

The fourth-order balanced filter was implemented in the same BiCMOS process. The filter was optimized for an operating center frequency of 150 MHz. The integrating capacitors were set to $2C_1 = 32$ pF and $2C_2 = 8$ pF, with 6 Ω and 12 Ω resistors in series for excess phase shift compensation. The chip was packaged in a standard PGA. For the test setup, the input and output terminals were connected to the signal generator and measuring instrument through 50 Ω cables, 50 Ω SMA connectors, and 50 Ω microstrip lines on the PCB.

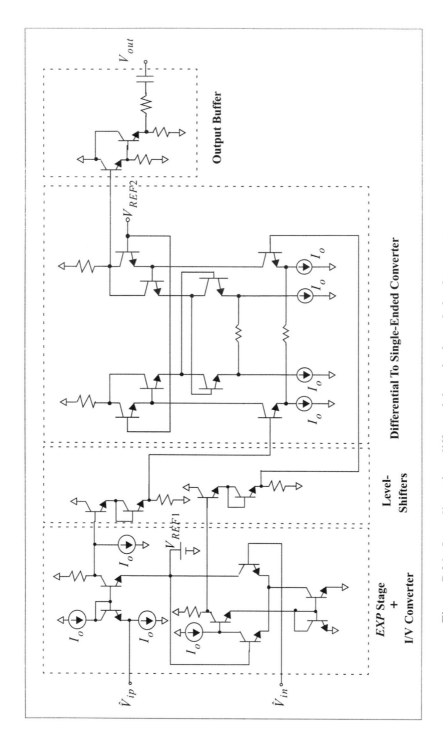

Figure 7-24: Low-distortion, differential-to-single-ended voltage output stage.

7.4.2 Frequency Performance

Figure 7-25 shows the measured as well as the simulated frequency responses for a bias current of 220 µA resulting in a center frequency of 130 MHz. At the lower end of the frequency range we notice a 40 dB/decade rising frequency response that is mainly due to the two 60 pF input and output decoupling capacitors. The response then tends to level-off at around 20 MHz due to the finite attenuation of the filter. This is mainly attributed to the mismatches between the various current sources. At the upper end of the frequency range, the filter's attenuation is higher than expected from simulation. It is believed that it is caused by the parasitic capacitors not accounted for during simulation, as well as to the finite gains of the CMFB circuits which resulted in slight shifts in the common-mode voltages at the capacitor nodes.

7.4.3 Linearity And Noise Measurements

In order to compare the distortion of this balanced implementation to that of the single-ended biquad discussed previously in Section 7.3, the center frequency of the filter was set to 75 MHz. This is approximately the same frequency as that used to measure the distortion of the biquad (Figure 7-21) in test chip#3. Two tones were applied at 74 MHz and 76 MHz such that their sum caused a current modulation index

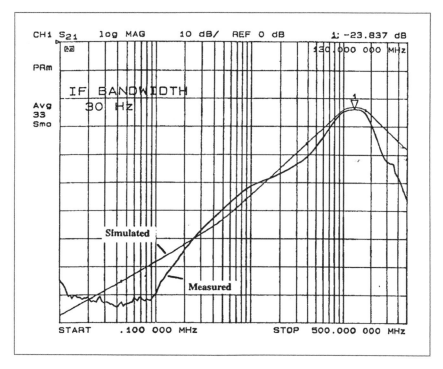

Figure 7-25: Measured and simulated frequency response of the fourth-order log-domain bandpass filter.

(I_{in}/I_{bias}) of 50% when biased at a current level of 121 μA. The spectrum of the output is given in Figure 7-26 showing an IMD3 of about -45.6 dB. Correspondingly, the IIP3 metric was found to be 0.58 mA.

Finally, the output noise power spectral density was measured to be -152.41 dBm/Hz (5.3 nVrms/$\sqrt{\text{Hz}}$) at the center frequency of 130 MHz. This is much lower than the noise of the previous biquad (146 nVrms/$\sqrt{\text{Hz}}$), further emphasizing the benefit of employing a balanced structure. The power consumption of the filter, including the input and output interface stages as well as the output buffer, was 343 mW for a center frequency of 130 MHz.

7.5 Test Chip #5: Low-Voltage 3rd-Order Chebyshev Filter

Our next experiment involves the high-speed (all-npn) low-voltage integrator of Section 2.3.3. Here we shall describe the measured performance of this integrator in the design of a third-order lowpass Chebyshev filter derived as an operational

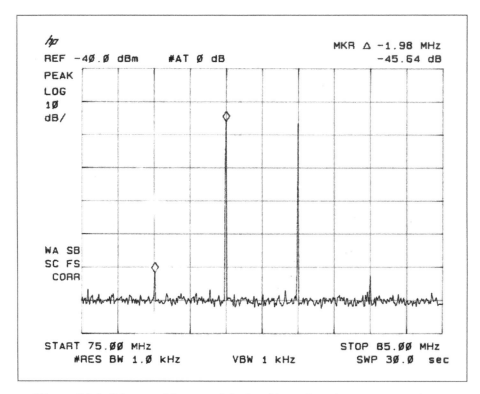

Figure 7-26: Measured intermodulation distortion of the balanced fourth-order filter.

simulation of an *LC* ladder prototype [46]. Although the integrator is expected to operate in the VHF band of 30 - 300 MHz, this particular realization was limited to 4 MHz operation at 1.2 V. This was due to the semi-custom implementation with off-chip capacitors of 10 pF and 22 pF. Nonetheless, the results provide a useful proof-of-concept for the low-voltage integrator.

7.5.1 The Test Set-Up

Unlike the previous designs, this particular implementation involved constructing a single integrator on a single chip using the GA911 semi-custom bipolar array technology from Gennum Inc.. Four separate cells, together with three capacitors, were assembled on a printed circuit board as to realize a third-order Chebyshev filter with 1.2 dB of passband ripple. In addition, input-output cells that perform (1) V-to-I or I-to-V operation, and (2) *LOG* or *EXP* operations were added to each chip. These cells are illustrated in Figure 7-27.

The input cell shown in Figure 7-27(a) consists of three main transistors: Q_1, Q_2 and Q_3, together with several current sources and voltage references for biasing. Biasing transistor Q_1 at a constant current I_o and setting its base voltage to a fixed voltage reference of 1.0 V, turns the emitter terminal of Q_1 into an AC virtual ground. A resistor R_{in} is connected between the input voltage node and the AC ground to perform the V-to-I operation. Transistor Q_2 then mirrors the input current over to Q_3 from which it serves as the *LOG*-input to the first log-domain integrator cell represented by transistors Q_4 and Q_5.

The output stage shown in Figure 7-27(b) consists of two transistors, some biasing circuits and a resistor R_{out}. A log-domain voltage is applied to the input of transistor Q_1 from which it is level-shifted upwards towards the positive power supply and applied to the base of Q_2. This voltage is then converted to a current via the exponential V-to-I behavior of Q_2 and the fact that its emitter terminal is held at a constant voltage of 0.3 V. Hence, Q_2 performs the necessary *EXP* operation. Subsequently, the linear output current from the collector of Q_2 is converted to a voltage via resistor R_{out}.

7.5.2 Frequency Performance

The filter was biased at a current level of 12.2 µA and the power supply was set to 1.2 V. The measured frequency response of the filter and its expected ideal behavior is shown in Figure 7-28. As evident, the cut-off frequency of the filter is slightly lower than the desired value. This is expected when one accounts for nonideal transistor behavior (see, for instance, the discussion in Chapters 5 and 6). By increasing the bias current to 13 µA one can easily correct for this frequency shift. The passband ripple and stopband attenuation behavior follows very closely with their expected values. Moreover, the filter was tunable from 40 kHz to 4 MHz when the

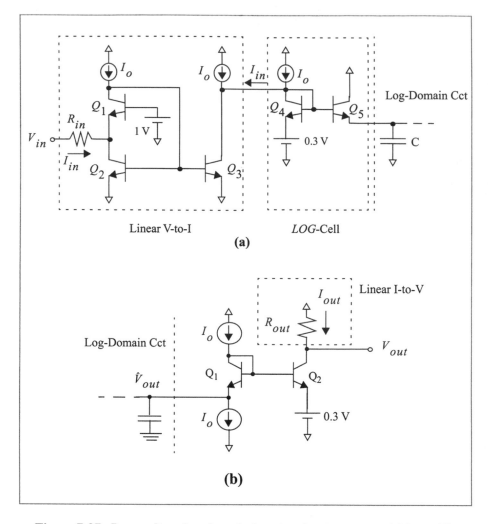

Figure 7-27: **Low-voltage log-domain input and output stage: (a) input V-to-I converter and *LOG* operation, (b) *EXP* operation and output I-to-V converter.**

bias current was varied from 1.2 μA to 76.5 μA.

7.5.3 Linearity and Noise Measurements

The filter was tuned to a cut-off frequency of 1.5 MHz using a bias current of 25 μA. The distortion of the filter was then measured using a one and two-tone test. The one-tone test was performed with a test frequency of 100 kHz such that the first 10 harmonics fell well within the filter's passband. The input amplitude was then varied until the best total harmonic distortion (THD) measure was obtained. This was

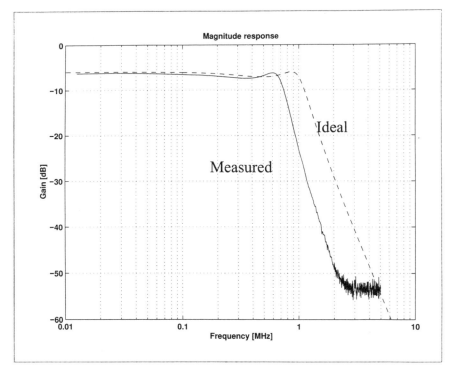

Figure 7-28: Measured and simulated frequency response of the low-voltage 3rd-order Chebyshev filter.

found to be -43.5 dB corresponding to an input current level of 1.4 μA. A two-tone test was also performed consisting of a tone at 498 kHz and 502 kHz, each having a current amplitude of 4 μA. The measured IMD3 was found to be -53.5 dB. Correspondingly, the IIP3 metric is 0.087 mA.

Lastly, the output noise power spectral density was measured to be 52.6 nA-rms over a 6 MHz bandwidth. The power consumption of the filter, including the input and output interface stages was 844 μW

7.6 Test Chip #6: Third-Order State-Space Filter

Following the principles laid down in Chapter 4, a third-order state-space log-domain filter [33] was fabricated using Nortel's 0.8 μm BiCMOS technology described previously. The filter is designed based on the high-speed (all-npn) log-domain integrator of Figure 2-7 in Section 2.3.1. Filter tunability is provided by an array of on-chip digitally-programmable dc (bias) current sources. Each current source consists of a simple eight-bit digital-to-analog converter (DAC) and an eight-bit shift register. It is capable of providing 256 distinct current levels to cover the wide range of bias currents typical of state-space filters. The shift registers of the current sources are daisy-chained together so that a single external bit stream could be used to control the current levels of each current source.

7.6.1 The Test Set-Up

For the most part, the test set-up was identical to that described for the test chip in Section 7.3 with the added feature of the digital control provided with a VXI-based digital generator. In addition, the V-to-I and I-to-V sections were implemented all on chip using a simple cell consisting of several differential pairs as shown in Figure 7-29. The simulated performance of the combined input and output stage limited the measurement range to about 50 dB. Figure 7-30 shows the microphotograph of the filter chip. It occupies an area of 1000 μm by 1000 μm.

Three 20 pF capacitors are implemented on-chip. A bias current of 200 μA is chosen so that it enables the filter to be tuned over a 2 to 8 MHz frequency range. The chip dissipates approximately 70 mW of power.

7.6.2 Demonstration of Filter Programmability

Figure 7-31 illustrates the behavior of the filter when programmed for a low-pass response with different cut-off frequencies, stop-band attenuations and passband gains. Part (a) depicts the filter's pass-band edge frequency tuned from 2 MHz to 8 MHz. As the frequency is increased, the notch becomes less defined, but remains visible. The stop-band attenuation of the filter can also be programmed over a wide range as shown in Fig. 7-31(b). Note that the two most attenuated cases reveal the

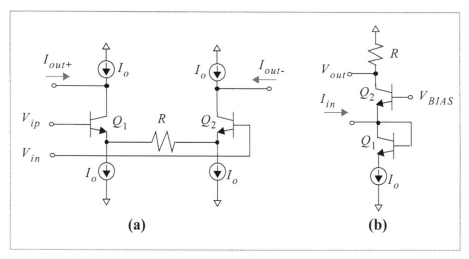

Figure 7-29: Simple input and output V-to-I and I-to-V stages: (a) input V-to-I converter (b) output I-to-V converter.

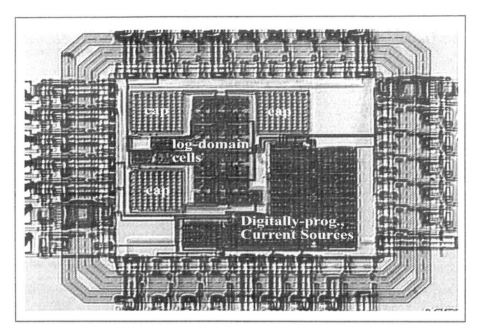

Figure 7-30: Microphotograph of the third-order state-space filter.

presence of a secondary zero. This secondary zero sets a limit to the maximum possible stop-band attenuation. The source of the zero is suspected to be associated with the test setup. Finally, the gain of the filter can also be controlled. An example of gain adjustment is shown in Fig. 7-31(c) where the gain is altered by about 25 dB.

Figure 7-32 illustrates how the filter can be coarsely tuned to give elliptic, Chebyshev or Butterworth responses with a pass-band attenuation of 0.5 dB. As is evident, the Butterworth pass-band behavior is slightly larger than expected. This is attributed to the lack of precision in the present tuning algorithm. Table 7-3 summarizes the characteristics of the third-order programmable log-domain filter.

7.7   Test Chip #7: Fifth-Order State-Space Filter

A second state-space filter realization was constructed, but this time based on a fifth-order formulation [74]. The design details are similar to the third-order realization of the previous subsection except that the filter was designed for balanced operation using the log-domain cells of Fig. 2-8. Figure 7-33 shows a floorplan and microphotograph of the filter chip. The chip was fabricated using a 0.8 μm BiCMOS technology and occupied an active area of 4000 μm by 3000 μm. It employed a nominal bias current of 5 mA for a maximum operating frequency of 50 MHz. Five interleaved arrays of 20 pF capacitors were implemented on-chip. The chip was digitally programmed to provide a 5th-order lowpass elliptic response (in particular,

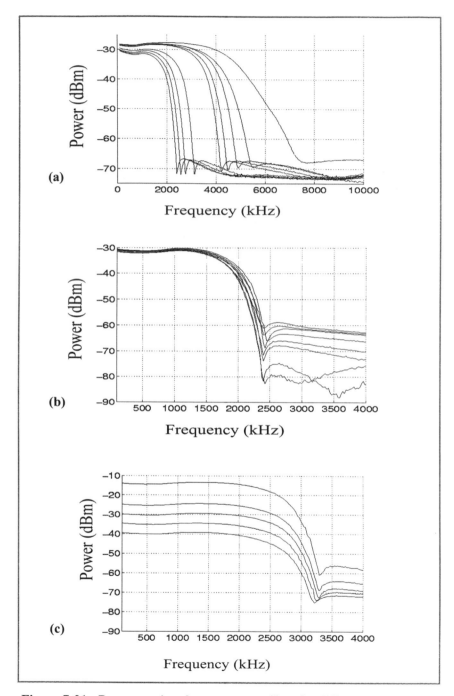

Figure 7-31: Programming the state-space filter for different (a) cutoff frequencies, (b) stopband attenuations, (c) passband gains.

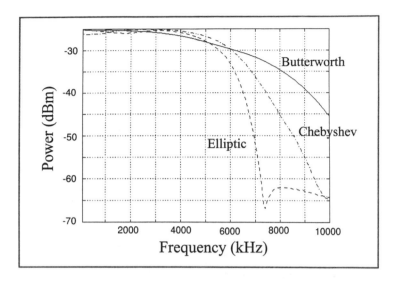

Figure 7-32: Tuning the shape of the state-space filter.

one that demonstrates the clear placement of several finite transmission zeros).

7.7.1 Measured Elliptical Behavior

The measured frequency response of the fifth-order state-space filter is shown in Figure 7-34. Here the filter has a passband ripple of about 0.5 dB, a 3 dB cut-off frequency of 20 MHz and a stopband attenuation of about 40 dB from 25 MHz to 50 MHz. In addition, one can clearly see the presence of two transmission zeros at 27 and 39 MHz. Table 7-4 summaries the measured results from this chip.

Die Size	1 mm^2
Supply Voltage	> 2.5 V
Power Consumption	70 mW
Maximum Operating Frequency	8 MHz
Adjustable Stop-band Attenuation	20 to 50 dB
THD (for input modulation level of 50%)	1.0%

Table 7-3: Summary of measured results for 3rd-order state-space filter

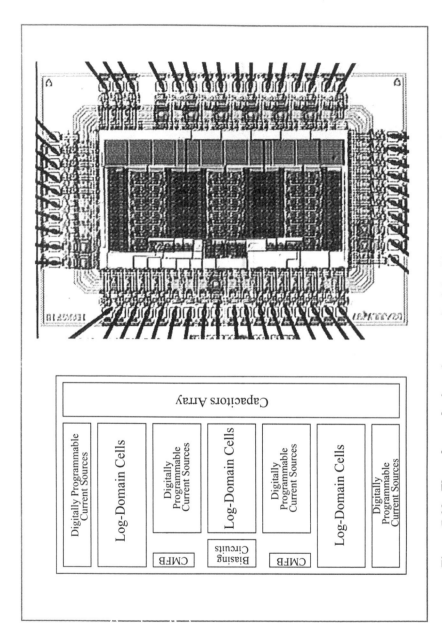

Figure 7-33: Floorplan and microphotograph of the fifth-order state-space filter.

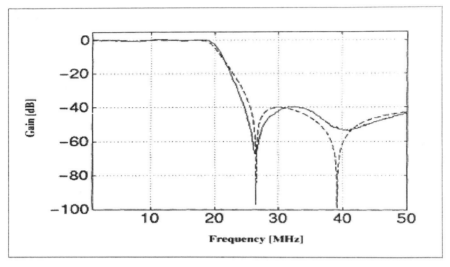

Figure 7-34: Measured (solid) versus simulated (dashed) frequency response for a fifth-order elliptic state-space filter.

7.8 Summary

In this chapter, we have presented the measurements of several log-domain filter chips. These results are of particular importance since they confirm the novel log-domain concept, from which linear filters are synthesized from nonlinear integrators. Also, issues regarding the practicality of these filters, such as silicon area, power consumption, linearity, noise level, and the interfaces (input V/I and output I/V) for current-mode signal processing were discussed.

Die Size	12 mm^2
Supply Voltage	> 2.5 V
Power Consumption	350 mW
Maximum Operating Frequency	50 MHz
Adjustable Stop-band Attenuation	20 to 40 dB
THD (for input modulation level of 50%)	0.5%

Table 7-4: Summary of measured results for 5th-order state-space filter.

The first two test chips (Sections 7.1 and 7.2) were among the first experimental results reported on this topic. They showed excellent correlation between the experimental results and their desired specifications. Electronic tunability has been achieved for at least two decades by simply adjusting the bias current level. Several distortion measurements were performed in order to verify their linearity. The harmonic distortion (THD) measured -62 dB for the biquad, and -47 dB for the fifth-order Chebyshev. Intermodulation distortion was -70 dB for the biquad and -55 dB for the Cheybyshev filter.

For high frequency operations, experimental results from a monolithic single-ended biquad (Section 7.3) and a fourth-order balanced bandpass filter were presented (Sections 7.4). They were designed based on the all-npn log-domain integrator circuits. A maximum operating frequency with a wide tuning range (≈ 140 MHz) and independent Q-tuning was demonstrated for the single-ended biquad. Low distortion and noise were achieved for the balanced fourth-order filter, which attained a maximum center frequency of 130 MHz. These results suggest the potential of log-domain filters in very high-frequency applications.

For low-voltage operation, a third-order Chebyshev filter was presented in Sections 7.5. It was designed based on the low-power high-speed integrator circuit (Section 2.3.3), and integrated in a semi-custom inexpensive bipolar process. Despite its semi-integration, the filter reached a maximum operating frequency of 4 MHz for $V_{cc} = 1.2$ V. Although single-ended and class A, the measured dynamic range was a respectable 41 dB.

Lastly, two test chips featuring highly programmable third-order and fifth-order state-space filters were presented (Sections 7.6 & 7.7). The results clearly demonstrate the electronic tunability of the filter's cutoff frequency, gain, as well as the pole/zero locations. Also based on the all-npn log-domain integrator circuit, the filter is capable of operating up to 8 MHz in the third-order case, and 50 MHz in the fifth-order case.

Appendix. A

Seevinck's Companding Current-Mode Integrator

Driven by the need for achieving high-speed filtering function under low supply voltages, Seevinck proposed a novel companding current-mode integration scheme in 1990 [47]. Interesting, a linear integrator is designed out of nonlinear parts. The input current signal is compressed, and integrated by a capacitor to form a voltage, which when expanded forms an output linear current. Since both input and output signals are in the form of current, and signal compression and expansion are involved, the circuit was termed a companding current-mode integrator.

Although its development was totally independent of the log-domain filters (in fact, the log-domain principles did not catch much attention until it was resurrected by Frey [22] a few years later), the underlying principles are strikingly similar. Therefore, for completeness, this integrator is reviewed here. At the end, we will point out its relationship to log-domain integrators.

The block diagram of the integrator is shown in Figure A-1(a). A divider combines the input and output current and produces a compressed signal $I_o(I_{in}/I_{out})$, where I_o is a normalizing constant[†]. The current is integrated in capacitor C, and produces a voltage v, which is given by

---

†. As we will see shortly, this constant is physically realized by a dc bias current.

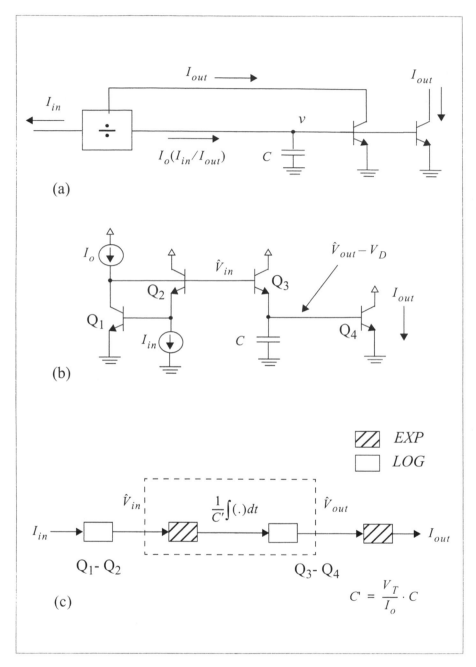

Figure A-1: Demonstration of companding current-mode integrator: (a) principle, (b) simplistic circuit implementation, and (c) log-domain SFG representation.

$$I_o \cdot \left(\frac{I_{in}}{I_{out}}\right) = C \cdot \frac{dv}{dt} \tag{A.1}$$

A bipolar transistor then serves as a transconductor to expand v into the current I_{out} with the familiar exponential relationship,

$$I_{out} = I_S e^{v/V_T} \tag{A.2}$$

Eliminate v by combining (A.1) and (A.2), we have

$$I_o \cdot \left(\frac{I_{in}}{I_{out}}\right) = C \cdot \frac{d}{dt}\left[V_T \cdot \ln\left(\frac{I_{out}}{I_S}\right)\right] \tag{A.3}$$

Integrating each side, and arrange, we arrive at

$$I_{out} = \frac{I_o}{V_T C} \cdot \int I_{in} dt \tag{A.4}$$

Although the intermediate steps involve nonlinear signal operations such as division and exponentiation, the input-output behavior is linear. Notice that the unity gain frequency is controllable through the constant I_o. Seeing that the currents in (A.1) are arranged in product pairs, by the translinear principle described in Chapter 1, the companding integrator can be realized by the simple circuit of Figure A-1(b)[†].

Now, we would like to show how the integrator circuit can be understood in using the log-domain analysis presented in Chapter 2. To do so, define the log-domain signals as shown in Figure A-1(b), and a pair of complementary LOG/ EXP operators as

$$LOG(X) = V_T \cdot \ln\left(\frac{I_o \cdot X}{I_S^2}\right) \qquad EXP(X) = \frac{I_S^2}{I_o} \cdot e^{X/V_T} \tag{A.5}$$

This then allows us to write the input log-domain voltage $\hat{V}_{in} = LOG(I_{in})$ generated by Q_1-Q_2 as

---

†. This circuit of Figure A-1(b) is not practical as it lacks a discharge path for the capacitor. However, it is sufficient to highlight the integrator concept without burdening the readers with non-essential details. For more complete and sophisticated implementations, such as the Class AB circuit, please refer to [47].

$$\hat{V}_{in} = V_D + V_T \cdot \ln\left(\frac{I_{in}}{I_S}\right)$$

$$= V_T \cdot \ln\left(\frac{I_o I_{in}}{I_S^2}\right)$$

(A.6)

where V_D is the diode drop across the bipolar transistor Q_1 with dc bias current I_o. Notice that $\hat{V}_{in}$ is equal to $LOG(I_{in})$. Transistor Q_3 and the capacitor then perform integration, and produce a level-shifted log-domain output voltage, $\hat{V}_{out} - V_D$. The KCL equation at the capacitor node is found to be

$$I_S \cdot e^{\frac{\hat{V}_{in} - \hat{V}_{out} + V_D}{V_T}} = C \cdot \frac{d}{dt}(\hat{V}_{out} - V_D)$$

(A.7)

Multiply both sides by $e^{\hat{V}_{out}/V_T}$, and rewrite V_D in terms of the diode equation, we have

$$I_o \cdot e^{\frac{\hat{V}_{in}}{V_T}} = C \cdot e^{\frac{\hat{V}_{out}}{V_T}} \cdot \frac{d}{dt}(\hat{V}_{out})$$

(A.8)

Apply the chain rule and re-arrange to obtain

$$\frac{I_S^2}{I_o} \cdot e^{\frac{\hat{V}_{in}}{V_T}} = \frac{CV_T}{I_o} \cdot \frac{d}{dt}\left(\frac{I_S^2}{I_o} \cdot e^{\frac{\hat{V}_{out}}{V_T}}\right)$$

(A.9)

In terms of the LOG operators, we arrive at a log-domain transfer function similar to the one in (2.5), given by

$$EXP(\hat{V}_{out}) = \frac{I_o}{CV_T} \cdot \int (EXP(\hat{V}_{in}))dt$$

(A.10)

Finally, transistor Q_4 acts as a exponential transconductor and converts the voltage $\hat{V}_{out} - V_D$ to a linear current I_{out}, so that

$$I_{out} = I_S \cdot e^{\frac{\hat{V}_{out} - V_D}{V_T}}$$

$$= \frac{I_S^2}{I_o} \cdot e^{\frac{\hat{V}_{out}}{V_T}} \tag{A.11}$$

$$= EXP(\hat{V}_{out})$$

The log-domain SFG is represented in Figure A-1(c). In principle, the companding current-mode integrator is remarkably similar to those used in log-domain circuits. Although the *LOG* and *EXP* operators presented here are different from those in (2.4), this circuit will share many characteristics of the log-domain integrator presented previously.

Appendix. B

Multitone Testing Using *SPICE*

SPICE simulation is limited by its inability to perform AC analysis on non-linear circuits. This is due to the fact that when performing AC analysis, *SPICE* first solves for the DC operating point of the circuit, then determines linearized, small-signal models for all of the non-linear devices in the circuit. This is particularly relevant when simulating the log-domain filter since the exponential nature of the bipolar transistor is at the very heart of operation.

The solution is to use a technique called *multitone testing* [71], [75]. Multitone testing involves applying a stimulus composed of one or more tones to the device under test and observing its spectral response. It is commonly used in the testing community and permits the simultaneous measurement of frequency response, total harmonic distortion (THD) and inter-modulation distortion (IMD), thereby reducing test time.

This appendix will show how SPICE transient analysis can be used to perform this kind of testing. The basic approach is as follows. First, SPICE is used to find the transient response of a circuit subjected to a multitone stimulus. Then the mathematical software package MATLAB finds the discrete Fourier transform of the transient output and plots it with respect to frequency.

B.1   Review of the Discrete Fourier Transform

In order to understand some of the constraints that will be imposed on the analysis, we begin with a review of the discrete Fourier transform (DFT) [76].

For every discrete-time sequence $x(n)$ there exists a Fourier transform which is defined by:

$$X(\omega) = \sum_{-\infty}^{\infty} x(n)e^{-j\omega n} \qquad\qquad (B.1)$$

Suppose we now limit the length of $x(n)$ to L samples, such that:

$$x'(n) = \begin{cases} x(n) & 0 \leq n \leq L-1 \\ 0 & n > L-1 \end{cases} \qquad\qquad (B.2)$$

The Fourier transform of this sequence will now be given by:

$$X(\omega) = \sum_{n=0}^{L-1} x'(n)e^{-j\omega n}$$

We now sample $X(\omega)$ at N equally spaced frequencies $\omega_k = \dfrac{2\pi k}{N}$, $k=0,1,...,N-1$ and $N \geq L$, such that:

$$X(k) \equiv X\!\left(\frac{2\pi k}{N}\right) = \sum_{n=0}^{L-1} x'(n)e^{-j\frac{2\pi kn}{N}}, \quad k = 0, 1, 2, ..., N-1$$

which can be rewritten as:

$$X(k) = \sum_{n=0}^{N-1} x'(n)e^{(-j2\pi kn)/N}, \qquad k = 0, 1, 2, ..., N-1 \qquad (B.3)$$

Equation (B.3) denotes the relationship for transforming a sequence $\{x'(n)\}$ of length $L \leq N$ into a sequence of frequency samples $\{X(k)\}$ of length N, and is called the *discrete Fourier transform* (DFT). The relationship that allows us to recover the sequence $\{x'(n)\}$ from a set of frequency samples is called the inverse discrete Fourier transform (IDFT) and is given by:

$$x'(n) = \frac{1}{N} \cdot \sum_{k=0}^{N-1} X(k)e^{(j2\pi kn)/N}, \qquad n = 0, 1, 2, ..., N-1 \qquad (B.4)$$

The DFT is usually computed using any of a number of efficient algorithms that are called Fast Fourier Transforms (FFT).

B.2   The *SPICE* File

The first step in multitone analysis is to use *SPICE* to find the transient response of a circuit stimulated by one or more sinusoidal tones. For those who are unfamiliar with transient analysis, *SPICE* finds the circuit solution at a set of discrete-time intervals. Consequently, the outcome of the analysis will be a sequence of discrete-time samples $\{x'(n)\}$ that can be used to compute the discrete Fourier transform as described in the previous section.

The *SPICE* file describes the different circuit components that make up the log-domain filter [77]. When preparing a *SPICE* file for multitone analysis, special care must be taken in two particular areas. First, the signal sources that make up the multitone input must be properly defined such that they have the right frequency, amplitude and phase. Each of these areas will be discussed at some length. Second, the transient analysis requests must be chosen correctly since the computation of the FFT depends on the proper transient output.

B.2.1   The Signal Sources

The circuit is to be stimulated by a set of sinusoidal current sources each operating at a different frequency. This section will outline some guidelines that govern the choice of frequency, amplitude and phase of the different tones in the multitone input.

B.2.1.1   Frequency

The choice of frequency for each different tone is important for two reasons:

1. The frequency of each input tone must correspond to one of the discrete frequency points in the DFT. This will ensure that every input tone completes an integral number of cycles over the total simulation time and thus prevents leakage and spreading effects.

2. The frequency of the different tones should not be multiples of one another. This will minimize the chance that their harmonics and intermodulation products coincide.

The choice of sampling frequency and of the unit test period (a concept to be defined next) will help us meet these constraints:

    a) The unit test period

In order to minimize leakage effects when calculating the DFT of a multitone system, we must ensure that <u>each</u> sinusoid in the multitone input completes an integral number periods over the time of the analysis. The shortest time interval which allows this for all tones is called the *unit test period* (UTP). For example, the unit test period for an input composed of 2 kHz, 3 kHz and 5 kHz sine waves is 1 ms, as shown in Figure B-1. The reciprocal of the UTP is called the primitive frequency and corresponds to the greatest divisor of the input frequencies (1 kHz in our example). The value of the primitive frequency will help determine the sampling frequency.

b) The sampling frequency

Since we want each of the input frequencies to correspond to one of the discrete frequencies in the DFT, we let the sampling frequency be given by:

$$f_s = N \cdot f_p \qquad \text{(B.5)}$$

where N = Number of points in the DFT
f_p = Primitive frequency

Figure B-2 shows the results of a 16 point DFT performed on a system composed of the three sine waves described in the previous section. Note that the sampling frequency is 16 kHz. Only the first eight samples are shown since the other eight are given by the mirror image of the first ones.

An additional problem that will affect the choice of the sampling frequency is aliasing. We know from the Nyquist Sampling Theory that in

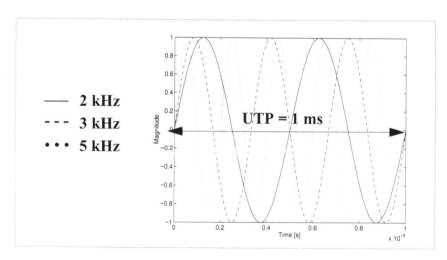

Figure B-1: Calculating the unit time period of a 3-tone input.

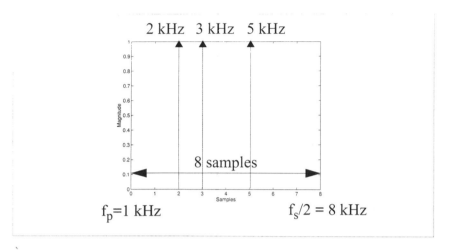

Figure B-2: Illustrating the relationship between the sampling frequency and other multitone parameters.

order to prevent aliasing, we must make sure that the highest frequency contained in the sampled signal is less than the Nyquist frequency, $f_s/2$. In other words, all of the input tones must be at frequencies less than $f_s/2$. Even with this precaution, we may get aliasing if any of the harmonics or the inter-modulation products are greater than the Nyquist frequency. In this case, some kind of anti-aliasing filter would be needed. Most of the analysis done is this work was done such that the cutoff frequency of the filter to be tested was much less than $f_s/2$. Therefore, the natural attenuation of the filter limits the effect of aliasing and eliminates the need for an anti-aliasing filter.

c) <u>Frequency resolution</u>

For a given sampling frequency, increasing the number of points in the DFT (*N*) will increase the frequency resolution of the spectral output. Changing *N* is analogous to controlling the resolution bandwidth on a spectrum analyzer. The drawback to choosing a large *N* is that the number of samples needed for the transient analysis is large and thus increases the simulation time. Usually, one attempts to find a balance between spectral resolution and run time. A second constraint on *N* is the type of FFT algorithm used. A radix-2 algorithm needs 2^N samples while a radix-3 algorithm needs 3^N samples and so on. In our case, the radix-2 algorithm was always used since it is generally faster. As a result, *N must always be a power of 2*.

d) Interference

Each tone must be placed at a frequency that will minimize the chance that its harmonics and intermodulation distortion terms fall in the same frequency bin. This is in general a very complicated process. Often one simply tries to use a scheme that forces any possible overlapping harmonics to be as high order as possible. There are a number of approaches that can be used:

Prime-Rich Signals

One strategy is to base the set of input frequencies exclusively on prime numbers. This ensures that none of the input frequencies are harmonic to one another and that no sum or difference products fall on a test frequency. Unfortunately, it does not prevent third and fifth order intermodulation interference.

An Iterative Scheme

Restricting the tones to prime numbers is a tedious process. Because of the large number of constraints which are placed on the multitone stimulus we eventually want to be able to automate the process by writing a computer program that will automatically give us the frequency and other parameters for each tone. As a result, a formula for deriving the frequency of each tone was borrowed from telecom CODEC applications. The frequency of each tone is computed as follows:

$$f_{tone_i} = \frac{M_i}{N} \cdot f_s \tag{B.6}$$

where:

$$f_s = sampling\ frequency$$

$$N = Number\ of\ points\ in\ the\ DFT$$

$$M_i = 9 + [i \times 16] \tag{B.7}$$

$$i = 0,1,2,...\ as\ long\ as\ M_i < (N/2)$$

This method is more susceptible to interference between the harmonics of the different input tones than the prime-rich scheme was. For example, the frequency of the 16th harmonic of the first input tone will be the same as the frequency of the 16th tone; hence, they will interfere with one another. The advantage of this particular scheme is that it is easier to implement algorithmically. The designer must decide whether the added interference is worth the greater ease of design.

B.2.1.2   Phase

As the periods of each sinusoid in the multitone input are related to one another, the sinusoids will tend to peak at the same points in time. When many tones are used this could lead to clipping in the circuit. The solution is to assign a random phase shift to each different sinusoid. This is easily done since *SPICE* can accept a phase shift as one of the parameters of the input.

B.2.1.3   Amplitude

Even with phase shifting, the average power of the multitone signal will increase with the number of tones used. To prevent overdriving the circuit, the amplitude of each tone will be restricted to some reasonable value. Our strategy is to assign the amplitude of each sinusoid according to the following equation:

$$\text{RMS of single tone} = \text{Desired RMS of multitone} / \sqrt{K} \qquad \text{(B.8)}$$

where K = the number of tones in the multitone signal. For a sinusoid,

$$\textit{Peak of single tone} = \textit{RMS of single tone} * \sqrt{2} \qquad \text{(B.9)}$$

Therefore,

$$\textit{Peak of single tone} = \textit{Desired RMS of multitone} * \sqrt{\frac{2}{K}} \qquad \text{(B.10)}$$

B.2.1.4   A Complete SPICE Current Source

A typical *SPICE* statement for one of the current sources in the multitone input is shown below:

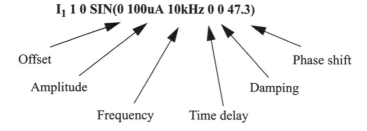

I₁ 1 0 SIN(0 100uA 10kHz 0 0 47.3)

Offset   Phase shift

     Amplitude   Damping

        Frequency   Time delay

Some of these parameters are of no consequence in our analysis and hence have been set to zero.

B.2.2   Analysis Requests

SPICE is told what type of analysis to perform by a command called an *analysis request*. We wish to perform transient analysis such that we produce a sequence of discrete-time samples $\{x'(n)\}$ that represent the output current of the log-domain filter. This sequence will then be used to compute its discrete Fourier transform

The *SPICE* request for transient analysis is done through the .TRAN command. A typical .TRAN statement is shown below:

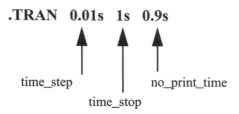

The three times are calculated as follows:
time_step: The step time is given by the sampling frequency.

$$time_{step} = \frac{1}{f_s} = T_s \tag{B.11}$$

time_stop: The stop time would normally be the time for one unit test period which is related to the sampling frequency by:

$$UTP = N \cdot T_s \tag{B.12}$$

However, we wish to allow the simulation to run long enough for the output to settle down [78]. Therefore, the total stop time is defined as:

$$time_{stop} = NP \cdot N \cdot T_s \tag{B.13}$$

where NP = *Number of periods needed for the simulation to settle down.*
no_print_time: This makes sure that only the last N points are printed, and is computed according to:

$$no\_print\_time = [(NP-1) \cdot N \cdot T_s] + T_s \tag{B.14}$$

We now have all the tools necessary to perform a spectral analysis of our circuit. The next section will show a simple example of how this would be done.

B.3   An Example using the Log-Domain Biquad

We wish to obtain the frequency response of the log-domain lowpass biquad (Figure 3-7) described in Chapter 3. As we have some flexibility in choosing the component values of the biquad, we will choose:

$$I_o = 100 \ uA$$

$$C_1 = 8.2 \ nF$$

$$C_2 = 2.2\ nF$$

which we know from theory should give us a cutoff frequency of:
$$f_c = 63.8\ kHz$$

The sampling criteria is chosen as follows:
$$f_s = 500\ kHz$$

$$N = 512\ points$$

Number of Tones =16

Number of Periods (NP) = 100

A C program was written which accounts for all of the criteria outlined in the previous two sections. The multitone inputs and the transient analysis statement were generated automatically and are shown below:

```
*** Multitone Sources ***
Is1 1 0 SIN(100uA 80uA  8789.0625 0 0
184.9)
Is2 1 0 SIN(100uA 80uA  24414.0625 0 0
63.2)
Is3 1 0 SIN(100uA 80uA  40039.0625 0 0
111.1)
Is4 1 0 SIN(100uA 80uA  55664.0625 0 0
192.4)
Is5 1 0 SIN(100uA 80uA  71289.0625 0 0
341.1)
Is6 1 0 SIN(100uA 80uA  86914.0625 0 0
61.8)
Is7 1 0 SIN(100uA 80uA  102539.0625 0 0
252.8)
Is8 1 0 SIN(100uA 80uA  118164.0625 0 0
81.5)
Is9 1 0 SIN(100uA 80uA  133789.0625 0 0
178.1)
Is10 1 0 SIN(100uA 80uA  149414.0625 0 0
44.8)
Is11 1 0 SIN(100uA 80uA  165039.0625 0 0
30.2)
Is12 1 0 SIN(100uA 80uA  180664.0625 0 0
140.2)
Is13 1 0 SIN(100uA 80uA  196289.0625 0 0
99.8)
Is14 1 0 SIN(100uA 80uA  211914.0625 0 0
132.5)
Is15 1 0 SIN(100uA 80uA  227539.0625 0 0
354.0)
```

```
Is16 1 0 SIN(100uA 80uA  243164.0625 0 0
192.7)

*** Analysis Requests ***
.TRAN 0.000002 0.1024 0.101378 0.000002
```

We then run *SPICE* and use *MATLAB* to calculate the FFT of the output. The result is plotted and is shown in Figure B-3.

The reader can clearly see the presence of the 16 tones of the multitone input. This allows us to verify the frequency response and measure the cutoff frequency, which is around 65 kHz, as expected. The smaller tones represent the many intermodulation products that occur when a multitone signal is applied to a circuit with non-linearity present in it. Overall, we can see how this offers an excellent alternative for finding frequency information about non-linear circuits. By varying the number of tones at the input we can get total harmonic distortion and intermodulation distortion results, as well as frequency response.

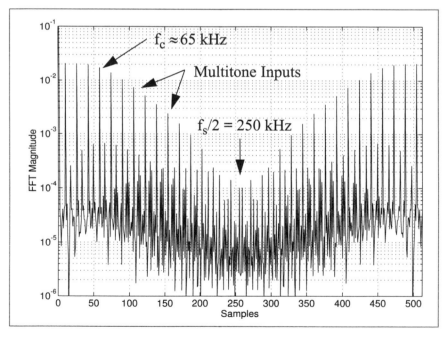

Figure B-3: FFT response of the log-domain biquad.

Appendix. C
Synthesis of High-Order Log-Domain Filters by a Cascade of Biquads

To illustrate the low-sensitivity of the ladder-based log-domain filter in Section 3.5 of Chapter 3, a seventh-order Chebyshev lowpass filter consisting of a cascade of biquads was used for comparison. In this appendix we shall describe the details of this filter.

A seventh-order filter consists of three second-order sections and one first-order section in cascade is shown in Figure C-1. The first three stages are similar to the lowpass biquad filter shown in Figure 3-7 without the output *EXP* stage. In addition, the input for each biquad is applied to the usually ground terminal of the *LOG* input stage. In this way, a natural *LOG-EXP* cancellation occurs between stages. The one-pole filter section is realized using the damped positive integrator shown in Figure 2-4 of Chapter 2. The overall input-output (I_{out}/I_{in}) linearity is maintained with the addition of an output *EXP* block.

For the Chebyshev filter approximation, a simple Matlab command [53] can be employed

```
[Z, P, K]= cheb1ap(N, Rp);
```

which returns the zeroes (Z), poles (P) and filter gain (K) when the filter order (N) and passband ripple (Rp, in dB) are specified. For a 7th-order filter, we can then group six

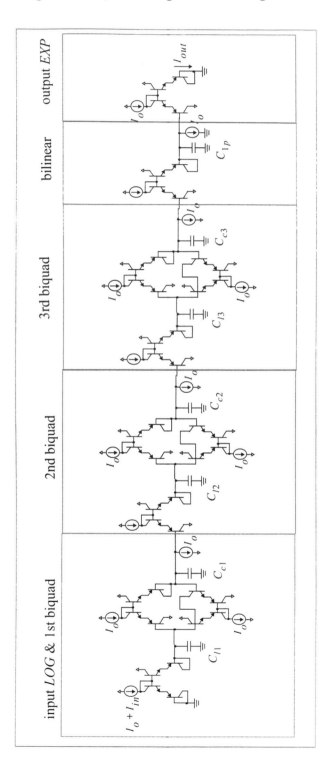

Figure C-1: 7th-order log-domain filter by biquad-cascade.

of the seven poles into three pairs. They will be realized by the three biquadratic filter sections. The remaining pole will be implemented by single first-order section. For a passband ripple of 1 dB and cutoff frequency of 1 MHz, the capacitor values of the log-domain filter were computed and are summarized in Table C-1.

1st biquad	C_{l1}	28.5155 pF	C_{c1}	3.3871 nF
2nd biquad	C_{l2}	0.1214 nF	C_{c2}	1.2089 nF
3rd biquad	C_{l3}	0.4973 nF	C_{c3}	0.8365 nF
bilinear	C_{1p}	1.5073 nF		

Table C-1: Capacitors of the 7th-order biquad-cascade log-domain filter.

Appendix. D

Derivation of Equation (5.16) - Compensation of Emitter Resistance, RE

In Chapter 5 we demonstrated a method in which to compensate for the effects of transistor nonzero emitter resistance RE. This involved tuning the log-domain integrator bias current from a nominal current level of I_o to a new value I_{comp} as described by (5.16). In this appendix, we shall derive this formula.

Assuming the bias current for the log-domain cell in Fig. 5-1 is changed from I_o to $k_c I_o$, we can write the RE-corrupted log-domain integrator expression as

$$k_c I_o e^{\frac{\hat{V}_{ip} - \hat{V}_o}{2V_T + 2R_E k_c I_o}} - k_c I_o e^{\frac{\hat{V}_{ip} - \hat{V}_o}{2V_T + 2R_E k_c I_o}} = C \frac{d}{dt} \hat{V}_o \tag{D.1}$$

which is derived simply by replacing each occurrence of I_o in (5.8) by $k_c I_o$. Rearranging the expression, we get

$$k_c I_o e^{\frac{\hat{V}_{ip}}{2V_T + 2R_E k_c I_o}} - k_c I_o e^{\frac{\hat{V}_{ip}}{2V_T + 2R_E k_c I_o}} = C \cdot e^{\frac{\hat{V}_o}{2V_T + 2R_E k_c I_o}} \frac{d}{dt} \hat{V}_o \tag{D.2}$$

which equals to

$$
k_c \left(I_o e^{\frac{\hat{V}_{ip}}{2V_T + 2R_E k_c I_o}} - I_o \right) - k_c \left(I_o e^{\frac{\hat{V}_{in}}{2V_T + 2R_E k_c I_o}} - I_o \right)
$$

$$
= \frac{2V_T C}{I_o} \cdot \left(\frac{2V_T + 2R_E k_c I_o}{2V_T} \right) \cdot \frac{d}{dt} \left(I_o e^{\frac{\hat{V}_o}{2V_T + 2R_E k_c I_o}} - I_o \right) \tag{D.3}
$$

Defining the *EXP* and *LOG* complementary mappings as

$$
EXP(x) = I_o e^{\frac{x}{2V_T + 2R_E k_c I_o}} - I_o
$$

$$
LOG(x) = (2V_T + 2R_E k_c I_o) \ln \left(\frac{I_o + x}{I_o} \right) \tag{D.4}
$$

we have

$$
k_c \cdot EXP(\hat{V}_{ip}) - k_c \cdot EXP(\hat{V}_{in}) = \frac{2V_T C}{I_o} \cdot \left(\frac{2V_T + 2R_E k_c I_o}{2V_T} \right) \cdot \frac{d}{dt} EXP(\hat{V}_o) \tag{D.5}
$$

Rearranging, the log-domain integrator equation becomes

$$
EXP(\hat{V}_o) = \left(\frac{V_T k_c}{V_T + R_E k_c I_o} \right) \cdot \frac{I_o}{2V_T C} \cdot \int \{EXP(\hat{V}_{ip}) - EXP(\hat{V}_{in})\} dt \tag{D.6}
$$

Comparing (D.6) and the ideal integration function (2.5), the effects of RE can be eliminated by setting the bracketed multiplicative factor to unity. Therefore, by setting

$$
\frac{V_T k_c}{V_T + R_E k_c I_o} = 1 \tag{D.7}
$$

the expression for k_c is found to be

$$
k_c = \frac{V_T}{V_T - R_E I_o} \tag{D.8}
$$

In summary, we can compensate for the RE nonideality by tuning I_o to $k_c I_o$ where k_c is given by (D.8). This is the result shown in equation (5.16).

Appendix. E

Goodness of Fit Test

We have assumed that deviations of the log-domain filter due to area mismatches are normally distributed. Based on this assumption, confidence intervals of the performance criteria are derived for any given σ_X^2 (variance of the I_S) in Section 5.4.1. Here we will inquire how good the filter deviations approximate normal distribution.

From the Monte-Carlo simulation, 1000 observations of the filter deviations are generated for $\sigma_X^2 = 10^{-4}$, 10^{-3} and 10^{-2}. The *goodness-of-fit* test is applied to determine whether that set of data may be looked upon as a random sample from a population having a normal distribution. Assuming the data has a normal distribution with an estimated mean and variance, the statistic

$$\chi^2 = \sum_{i=1}^{m} \frac{(f_i - e_i)^2}{e_i} \qquad \begin{aligned} & m = \text{number of partitions} \\ & \text{where } f_i = \text{observed frequency in each partition} \\ & e_i = \text{expected frequency in each partition} \end{aligned} \qquad \text{(E.1)}$$

should have a chi-square distribution with twenty-two degrees of freedom (i.e., m=25 and two estimated parameters in our assumed distribution) [62]. If the calculated χ^2 has low probability of occurrence, it implies the assumption of normal distribution is not appropriate. As is common statistical practice, a probability of 0.05 is considered "low".

Applying the above test to our simulated data set in Section 5.4.1, the results are tabulated in Table E-1.

σ_X^2	Probability of the calculated χ^2		
	f_{oa}/f_o	Q_a/Q	K
1e-4	0.316	0.297	0.903
1e-3	0.724	0.762	0.196
1e-2	0.194	0.106	< 0.0001

Table E-1: Goodness-of-fit test on the simulated data shown in Section 5.4.1.

As displayed above, most of the probabilities take on values much higher than 0.05, showing that the normal-distribution-fit is very good and valid. Only when σ_X^2 becomes large (i.e., the variance of the transistor area is large), the normal distribution assumption becomes marginal and unacceptable.

References

[1]   Y. P. Tsividis and J. O. Voorman, Eds., *Integrated Continuous-Time Filters*, Piscataway, NJ: IEEE Press, 1993.

[2]   A. S. Sedra and K. C. Smith, *Microelectronics Circuits, 3rd Edition*, Florida, USA: Saunders College Publishing.

[3]   A. Sedra and P. Brackett, *Filter Theory and Design: Active and Passive*, Portland, OR: Matrix Publishers Inc., 1978.

[4]   S. Sakurai, M. Ismail, J. Michael, E. Sanchez-Sinencio and R. Brannen, "A MOS-C variable equalizer with simple on-chip automatic tuning," *IEEE J. of Solid-State Circuits*, vol. 27, no. 6, pp. 927-034, June 1992.

[5]   J. van der Plaus, "MOSFET-C filter with low excess noise and accurate automatic tuning," *IEEE J. of Solid-State Circuits*, vol. 26, no. 7, pp. 922-929, July 1991.

[6]   M. Banu and Y. P. Tsividis, "An elliptic continuous-time CMOS filter with on-chip automatic tuning," *IEEE J. of Solid-State Circuits*, vol. 20, no. 6, pp. 1114-1121, Dec. 1985.

[7]   U. Moon and B. Song, "A low-distortion 22 kHz 5th-order Bessel filter," *IEEE Solid-State Circuits Conf., Dig. Tech. Papers*, vol. 36, pp. 110-111, Feb. 1993.

[8]   A. B. Grebene, *Bipolar and MOS Analog Integrated Circuit Design*, New York, NY: John Wiley and Sons, 1984.

[9]   W. M. Snelgrove and A. Shoval, "A balanced 0.9 μ m CMOS transconductance-C filter tunable over the VHF range," *IEEE J. of Solid-State Circuits*, vol. 27, no. 3, pp. 314-323, March 1992.

[10]   B. Nauta, "A CMOS transconductance-C filter technique for very high frequencies," *IEEE J. of Solid-State Circuits*, vol. 27, no. 2, pp. 142-153, Feb. 1992.

[11]   M. I. Ali, M. Howe, E. Sanchez-Sinencio and J. Ramirez-Anulo, "A BiCMOS low distortion tunable OTA for continuous-time filters", *IEEE Trans. on Circuits and Systems I*, vol. 40, no. 1, pp. 43-49, Jan. 1993.

[12]   J. Silva-Martinez, M. Steyaert and W. Sansen, "A large-signal very low-distortion transconductance for high-frequency continuous-time filters," *IEEE J. of Solid-State Circuits*, vol. 26, pp. 946-955, July 1991.

[13]   S. D. Willingham and K. W. Martin, "A BiCMOS low-distortion 8 MHz lowpass filter, *IEEE Solid-State Circuits Conf., Dig. Tech. Papers*, vol. 36, pp. 114-115, Feb. 1993.

[14]   R. Schaumann and M. A Tan, "The problem of on-chip automatic tuning in continuous-time integrated filters," *Proc. IEEE Int. Symp. on Circuits and Systems*, pp. 106-109, Feb. 1989.

[15]   K. A. Kozma, D. A. Johns and A. S. Sedra, "Tuning of continuous-time filters in the presence of parasitic poles," *IEEE Trans. on Circuits and Systems I*, vol. 40, no. 1, pp. 13-20, Jan. 1993.

[16]   B. Gilbert, "Current mode circuits from a translinear viewpoint: a tutorial," *Analog IC Design: The Current-Mode Approach*, C. Toumazou. F. Lidgey, and D. Haigh (eds.), IEE Circuits and Systems Series, vol. 2. Peter Perigrious Press, pp. 11-92, 1990.

[17]   D. Frey, "Log-domain filtering for RF applications," *IEEE J. of Solid-State Circuits*, vol. 31, no. 10, pp. 1468-1475, Oct. 1996.

[18]   M. El-Gamal, V. Leung and G. W. Roberts, "Balanced log-domain filters for VHF applications," *Proc. IEEE Int. Symp. on Circuits & Systems*, pp. 493-496, June 1997.

[19]   F. Yang, C. Enz and G. van Ruymbeke, "Design of low-power and low-voltage log-domain filters," *Proc. IEEE Int. Symp. on Circuits & Systems*, pp. 117-120, May 1996.

[20]   Y. Tsividis, "Externally linear, time-invariant systems and their application to companding signal processors," *IEEE Trans. on Circuits & Systems II*, vol. 44, no. 2, pp. 65-85, Feb. 1997.

[21]   R. W. Adams, "Filtering in the log-domain," *Preprint #1470 presented at the 63rd AES Conference*, New York, NY, May 1979.

[22]   D. Frey, "Log-domain filtering: an approach to current mode filtering," *IEE Proceedings-G*, vol. 140, no. 6, pp. 406-416, Dec. 1993.

[23]   D. Frey, "A general class of current mode filters," *Proc. IEEE Int. Symp. on Circuits & Systems*, pp. 1435-1438, May 1993.

[24]   D. Frey, "Current mode class AB second order filter," Electronics Letters, vol. 30, no. 3, pp. 205-206, Feb. 1994.

[25]   D. Frey, "A 3.3 Volt electronically tunable active filter usable to beyond 1 GHz," *Proc. IEEE Int. Symp. on Circuits & Systems*, pp. 493-496, May-June 1994.

[26]   D. Frey, "Exponential state space filters: A generic current mode design strategy," *IEEE Trans. Circuits and Systems I*, vol. 43, no.1 pp. 34-42, Jan. 1996.

[27]   D. Frey, "Log filtering using gyrators", Electronics Letters, vol. 32, no. 1, pp. 26-28, Jan. 1996.

[28]   D. Frey, "An adaptive analog notch filter using log filtering," *Proc. IEEE Int. Symp. on Circuits & Systems*, pp. 297-300, May 1996.

[29]   D. Perry and G. W. Roberts, "Log-domain filters based on LC ladder synthesis," *Proc. IEEE Int. Symp. on Circuits & Systems*, pp. 311-314, May 1995.

[30]   D. Perry and G. W. Roberts, "Design of log-domain filters based on the operational simulation of LC ladders," *IEEE Trans. on Circuits & Systems II*, vol. 43, no. 11, pp. 763-773, Nov. 1996.

[31]   M. El-Gamal and G. W. Roberts, "*LC* ladder-based synthesis of log-domain bandpass filters," *Proc. IEEE Int. Symp. Circuits & Systems*, pp. 105-108, June 1997.

[32]   A. Hematy and G. Roberts, "A fully-programmable analog log-domain filter circuit," *Proc. 2nd IEEE-CAS Region 8 Workshop on Analog and Mixed IC Design*, Baveno, Italy, Sep. 1997.

[33]   A. Hematy and G. Roberts, "A fully-programmable analog log-domain filter circuit," *Proc. IEEE Int. Symp. of Circuits & Systems*, pp. 309-312, May 1998.

[34]   B. Gilbert, "Translinear circuits: a proposed classification," *Electron. Lett.*, vol. 11, no. 1, pp. 14-16, Jan. 1975.

[35]   B. Gilbert, "A new wideband amplifier technique," *IEEE J. of Solid-State Circuits*, SC-3(4), pp. 355-365, 1968.

[36]   B. Gilbert, "A precise four-quadrant multiplier with sub-nanosecond response," *IEEE J. of Solid-State Circuits*, SC-3(4), pp. 365-373, 1968.

[37]   B. Gilbert, "Translinear circuits- 25 years on," Part I-III, *Electronics Engineering*, Aug.- Oct., 1993.

[38]   B. Gilbert, "Translinear circuits: an historical overview," *Analog Integrated Circuits and Signal Processing*, vol. 9, n. 2, pp. 95-118, March, 1996.

[39]   Members of the Technical Staff of Bell Telephone Laboratories, *Transmission Systems for Communications*, Bell Telephone Laboratories, Inc., 5th ed., 1982.

[40]   Y. Tsividis, "Developments in integrated continuous-time filters," *Analog Circuit Design: Low-Power Low-Voltage, Integrated Filters and Smart Power*, R. J. van de Plassche, W. M. C. Sansen and J. H. Huijsing (eds.), pp. 129-148, Kluwer Academic Publishers, 1995.

[41]   E. M. Drakakis, A. J. Payne, and C. Toumazou, "Log-domain filters, translinear circuits, and the Bernoulli cell," *Proc. IEEE Int. Symp. Circuits & Systems*, pp. 501-504, June 1997.

[42]   M. N. El-Gamal and G. W. Roberts, "Very high-frequency log-domain bandpass filters," *IEEE Trans. on Circuits and Systems II*, vol. 45, no. 9, pp. 1188-1198, Sept. 1998.

[43]   M. Punzenberger and C. Enz, "A new 1.2V BiCMOS log-domain integrator for companding current-mode filters," *Proc. IEEE Int. Symp. Circuits & Systems*, pp. 125-128, May 1996.

[44]   M. Punzenberger and C. Enz, "A compact low-power BiCMOS log-domain filter," *IEEE J. of Solid State Circuits*, vol. 33, no. 7, pp. 1123-1129, July 1998.

[45]   M. Punzenberger and C. Enz, "A 1.2-V low-power BiCMOS class AB log-domain filter," *IEEE. J. Solid State Circuits*, vol. 32, no. 12, Dec. 1997.

[46]   M. N. El-Gamal and G. W. Roberts, "A new 1.2 V npn-only log-domain integrator," *Proc. Int. Symp. Circuits & Systems*, pp. II-681 - II-684, June 1999.

[47]   E. Seevinck, "Companding current-mode integrator: a new circuit principle for continuous-time monolithic filters," *Electronics Letters*, vol. 26, no. 24, pp. 2046-2047, Nov. 1990.

[48]   M. Van Valkenburg, *Analog Filter Design*, New York, NY:, Holt, Rinehart and Winston, 1982.

[49]   W. Snelgrove and A. Sedra, *FILTOR2- A Computer Aided Filter Design Package*, Champagne, Ill.: Matrix Publishers.

[50]   C. Ouslis, W. M. Snelgrove and A. S. Sedra, "A filter designer's filter design aid: filtorX," *Proc. IEEE Int. Symp. on Circuits & Systems*, pp. 376-379, 1991.

[51]   J. Vlach and K. Singhal, *Computer Methods for Circuit Analysis and Design*, New York, NY: Van Nostrand Reinhold, 1983.

[52]   Gennum Corporation, *Gennum Data Book*, 1995.

[53]   The MathWorks Inc., *MATLAB Reference Guide*, 1995.

[54]   W. M. Snelgrove and A. S. Sedra, "Synthesis and analysis of state-space active filters using intermediate transfer functions," *IEEE Trans. on Circuits and Systems*, vol. CAS-33, pp. 287-301, Mar. 1986.

[55]   H. J. Orchard, "Inductorless filters," *Electronics Letters*, vol 2, pp. 224-225, June 1966.

[56]   H. J. Orchard, G. C. Temes, and T. Cataltyepe, "Sensitivity formulas for terminated lossless two-ports," *IEEE Trans. on Circuits and Systems*, vol. CAS-32, pp. 459-466, May 1985.

[57]   D. A. Johns, A. S. Sedra, "State-space Simulation of LC Ladder Filter," *IEEE Trans. on Circuits and Systems*, vol. CAS-34, pp. 986-988, August 1987.

[58]   V. Leung, M. El-Gamal and G. Roberts, "Effects of transistor nonidealities on log-domain filters," *Proc. IEEE Int. Symp. on Circuits & Systems*, pp. 109-112, June 1997.

[59]   V. Leung and G. W. Roberts, "Analysis and Compensation of Log-Domain Filter Response Deviations due to Transistor Nonidealities," *Analog Integrated Circuits and Signal Processing*, (in press).

[60]   E. Seevinck, *Analysis and Synthesis of Translinear Integrated Circuits*, Studies in Electrical and Electronic Engineering, Elsevier, 1988.

[61]   M. Ismail and T. Fiez (eds.), *Analog VLSI: Signal and Information Processing*, pp. 618-619, McGraw-Hill, 1994.

[62]   J. E. Freund and R. E. Walpole, *Mathematical Statistics (4th ed.)*, Englewood Cliffs, NJ: Prentice-Hall, 1987.

[63]   G. W. Roberts, "Gm-C integrator: magnitude and phase errors", *Analog Signal Processing Course Notes*, McGill University, 1998.

[64]   G. Temes and H. Orchard, "First order sensitivity and worst case analysis of doubly terminated reactance two-ports," *IEEE Trans. on Circuit Theory*, vol. CT-20, pp. 567-571, Sept. 1973.

[65]   M. Blostein, "Sensitivity analysis of parasitic effects in resistance terminated LC filters," *IEEE Trans. on Circuit Theory*, vol. CT-14, pp. 21-25, March 1967.

[66]   D. Perry, "The design of log-domain filters based on the operational simulation of LC ladders," *Master's Dissertation*, McGill University, Montreal, Canada, 1995.

[67]   R. A. Witte, *Spectrum and Network Measurements*, NJ: Prentice-Hall, 1991.

[68]   R. Caprio, "Precision differential voltage-current converter," *Electronics. Letters*, vol. 9, no. 6, pp. 147-148, Mar. 1973.

[69]   J. F. Jensen, G. Raghavan, A. E. Cosand, and R. H. Walden, "A 3.2GHz second-order delta-sigma modulator implemented in InP HBT technology," *IEEE J. Solid-State Circuits*, vol. 30, pp. 1119-1127, Oct. 1995.

[70]   M. El-Gamel, "Generalized log-domain integrator structure & its application to the synthesis of high-frequency and low-voltage log-domain filters," *Ph.D.'s Dissertation*, McGill University, 1998.

[71]   G. W. Roberts, "Calculating distortion levels in sampled-data circuits using SPICE," *IEEE Int. Symp. on Circuits and Systems*, Seattle, Washington, vol. 3, pp. 2059-2062, May 1995.

[72]   P. R. Belanger, *Control Engineering: A Modern Approach*, Oxford University Press, 1995.

[73]   Robert M. Fox; "Design-oriented analysis of log-domain circuits," *IEEE Trans. Circuit Syst. II*, vol. 45, pp. 918-921, July 1998.

[74]   A. Hematy, "Digitally programmable analog log-domain filters," *M. Eng.'s Dissertation*, McGill University, November 1998.

[75]   M. Mahoney, *DSP-Based Testing of Analog and Mixed-Signal Circuits*, New York, NY: IEEE Computer Society Press, 1987.

[76]   J. G. Proakis and D. G. Manolakis, *Digital Signal Processing*, New York, NY: MacMillan Publishing Co., 1992.

[77]   G. W. Roberts and A. S. Sedra, *SPICE, 2nd Edition*, Oxford University Press, 1992.

[78]   P. Crawley and G. W. Roberts, "Predicting harmonic distortion in switched-current memory circuits", *IEEE Trans. on Circuits and Systems -I: Fundamental Theory and Applications,* Vol. 41, pp. 73-86, Feb. 1994.

Index